W0079837

Telomerase Inhibition

METHODS IN MOLECULAR BIOLOGY™

John M. Walker, SERIES EDITOR

METHODS IN MOLECULAR BIOLOGY™

Telomerase Inhibition

Strategies and Protocols

Edited by

Lucy G. Andrews

Department of Biology, University of Alabama at Birmingham, Birmingham, AL

and

Trygve O. Tollefsbol

Department of Biology, University of Alabama at Birmingham, Birmingham, AL

HUMANA PRESS ✳ TOTOWA, NEW JERSEY

Production Editor: Christina M. Thomas

Cover design by: Karen Schulz

For additional copies, pricing for bulk purchases, and/or information about other Humana titles, contact Humana at the above address or at any of the following numbers: Tel.: 973-256-1699; Fax: 973-256-8341; E-mail: humana@humanapr.com; or visit our Website: www.humanapress.com

Printed in the United States of America. 10 9 8 7 6 5 4 3 2 1

Library of Congress Control Number: 2007933147

Preface

Telomerase, the enzyme that synthesizes telomeres, contributes to cellular immortality that is a basic feature of cancer cells. About 90% of cancer cells have robust telomerase activity, and this enzyme is expressed at very low levels in normal cells. The inhibition of telomerase or its associated proteins should selectively target cancer cell death with little or no effect on normal cells.

Due in part to the selective nature of telomerase inhibition as an anticancer approach, this field has expanded considerably from about 1998 to the present. The recent advances in methods of telomerase inhibition encompass many different areas of research including molecular biology, cell biology, biochemistry, oncology, and gerontology.

The techniques described in this book should provide the researcher with a diverse and comprehensive set of tools with which to study telomerase inhibition. Recently developed methods that have widespread application such as targeting the telomerase holoenzyme, its RNA template, and other elements associated with telomerase activity are presented. Additional methods involving the screening of telomerase inhibitors and telomerase inhibition combined with other chemotherapeutic agents are presented. This book should provide investigators with the most recent methods applied to the expanding field of telomerase inhibition.

Lucy G. Andrews
Trygve O. Tollefsbol

Contents

Contributors

LUCY G. ANDREWS • *Department of Biology, University of Alabama at Birmingham, Birmingham, AL*

AMANDA P. CUNNINGHAM • *Department of Biology, University of Alabama at Birmingham, Birmingham, AL*

HESHAM EL DALY • *Department of Hematology/Oncology, Freiburg Medical University Center, Freiburg, Germany*

SUSANNE FUESSEL • *Department of Urology, Technical University Dresden, Dresden, Germany*

YI-YUAN HUANG • *Institute of Biopharmaceutical Science, National Yang-Ming University, Taipei, Taiwan*

MOHAMMED KASHANI-SABET • *UCSF Comprehensive Cancer Center, San Francisco, CA*

INDZI KATIK • *Department of Immunology, Monash University Medical School, Melbourne, Australia*

ELKE KLEIDEITER • *Dr. Margarete Fiscer-Bosch-Institute of Clinical Pharmacology, Stuttgart, Germany*

ULRICH KLOTZ • *Dr. Margarete Fiscer-Bosch-Institute of Clinical Pharmacology, Stuttgart, Germany*

SEIJI KONDO • *Department of Neurosurgery, Anderson Cancer Center, Houston, TX*

YASUKO KONDO • *Department of Neurosurgery, Anderson Cancer Center, Houston, TX*

KAI KRAEMER • *Department of Urology, Technical University Dresden, Dresden, Germany*

SERENE R. LAI • *Department of Biology, University of Alabama at Birmingham, Birmingham, AL*

HE LI • *Centre for Functional Genomics and Human Disease, Monash University, Melbourne, Australia*

SHANG LI • *UCSF Comprehensive Cancer Center, San Francisco, CA*

JING-JER LIN • *Institute of Biopharmaceutical Science, National Yang-Ming University, Taipei, Taiwan*

JUN-PING LIU • *Department of Immunology, Monash University Medical School, Melbourne, Australia*

UWE M. MARTENS • *Department of Hematology/Oncology, Freiburg Medical University Center, Freiburg, Germany*

AXEL MEYE • *Department of Urology, Technical University Dresden, Dresden, Germany*

MEHDI NOSRATI • *UCSF Comprehensive Cancer Center, San Francisco, CA*

TOMOKAZU OHISHI • *Division of Molecular Biotherapy, Cancer Chemotherapy Center, Japanese Foundation for Cancer Research, Tokyo, Japan*

JUNKO H. OHYASHIKI • *First Department of Internal Medicine, Tokyo, Japan*

KAZUMA OHYASHIKI • *First Department of Internal Medicine, Tokyo, Japan*

KAMILLA PIOTROWSKA • *Dr. Margarete Fiscer-Bosch-Institute of Clinical Pharmacology, Stuttgart, Germany*

HIROYUKI SEIMIYA • *Division of Molecular Biotherapy, Cancer Chemotherapy Center, Japanese Foundation for Cancer Research, Tokyo, Japan*

JING-WEN SHIH • *Institute of Biopharmaceutical Science, National Yang-Ming University, Taipei, Taiwan*

TETSUZO TAUCHI • *First Department of Internal Medicine, Tokyo, Japan*

TRYGVE O. TOLLEFSBOL • *Department of Biology, University of Alabama at Birmingham, Birmingham, AL*

TAKASHI TSURUO • *Division of Molecular Biotherapy, Cancer Chemotherapy Center, Japanese Foundation for Cancer Research, Tokyo, Japan*

DAKANG XU • *Department of Immunology, Monash Medical School, Melbourne, Australia*

1

Methods of Telomerase Inhibition

Lucy G. Andrews and Trygve O. Tollefsbol

Summary

Telomerase is central to cellular immortality and is a key component of most cancer cells although this enzyme is rarely expressed to significant levels in normal cells. Therefore, the inhibition of telomerase has garnered considerable attention as a possible anticancer approach. Many of the methods applied to telomerase inhibition focus on either of the two major components of the ribonucleoprotein holoenzyme, that is, the telomerase reverse transcriptase (TERT) catalytic subunit or the telomerase RNA (TR) component. Other protocols have been developed to target the proteins, such as tankyrase, that are associated with telomerase at the ends of chromosomes. This chapter summarizes some of these recent advances in telomerase inhibition.

Key Words: Telomerase; inhibition; technique; method; telomerase reverse transcriptase (TERT); telomerase RNA (TR).

1. Introduction

Telomeres are sequences of DNA extending for many kilobases at the ends of chromosomes that in humans consist of hexameric 5′-TTAGGG-3′ tandem repeats. Telomerase is a ribonucleoprotein that maintains the lengths of chromosomal ends by synthesizing telomeric sequences. There are two major components of the telomerase holoenzyme: the telomerase reverse transcriptase (TERT) protein subunit that catalyzes the enzymatic reaction of DNA synthesis and the telomerase RNA (TR) component that serves as a template for the addition of deoxyribonucleotides to the ends of chromosomes. Although other

From: *Methods in Molecular Biology, vol. 405: Telomerase Inhibition*
Edited by: L. G. Andrews and T. O. Tollefsbol © Humana Press Inc., Totowa, NJ

proteins are associated with the holoenzyme, these two components are essential and sufficient for telomerase activity and telomere lengthening *(1–3)*.

2. Importance of Telomerase in Cancer and Aging

Telomerase is expressed in germline and embryonic stem cells as well as most somatic stem cells but is barely detectable in the great majority of adult somatic cells *(4)*. In actively dividing somatic cells, the telomeres shorten with each cell replication because of the paucity of telomerase activity eventually leading to replicative senescence after about 50 divisions *(5)*. This phenomenon has led to the idea that reduced telomerase expression in somatic cells may set in motion a molecular clock that controls the aging process *(5)*. In contrast, telomerase is up-regulated in the vast majority of cancer cells that are dependent upon this enzyme for maintaining their telomere lengths thereby conferring unlimited proliferative capacity or cellular immortality *(4)*.

Because telomerase is active in most cancer cells and is almost undetectable in most normal somatic cells, analysis of telomerase activity has potential as a diagnostic marker of cancer. The increase in telomerase in cancer cells generally occurs very early during tumorigenesis and sensitive techniques such as the telomeric repeat amplification protocol (TRAP) assay can detect trace levels of this enzyme, a method that has obvious potential for cancer diagnosis *(6)*.

3. Potential of Telomerase Inhibition in Cancer Therapy

The cancer therapeutic potential of telomerase inhibition is probably the area of telomerase research that has received the most attention. Telomerase activity can be inhibited in cancer cells and leads to a marked reduction in cellular viability and/or induces apoptosis of these cells *(7,8)*. Anticancer approaches directed at telomerase inhibition are varied, and methods ranging from RNA interference (RNAi) of the TERT catalytic subunit to inhibition of the proteins associated with telomerase at the telomeres have proven to have efficacy against cancer. As the telomeres of the rare normal cells that express telomerase are longer than that in most cancer cells and the level of telomerase activity is generally lower in normal telomerase-positive cells as compared with cancer cells, the risks associated with possible telomere shortening in normal cells because of off-target telomere shortening are thought to be relatively minimal. Therefore, the efficacy of telomerase inhibition in inducing loss of viability or apoptosis of cancer cells combined with the relative low risks to normal cells of inhibition of telomerase have moved telomerase research to the forefront among anticancer approaches.

4. Contents of this Book

4.1. Methods of Inhibiting the TERT Catalytic Subunit of Telomerase

The TERT catalytic subunit has been a major target for the development of anticancer methods because of the high concentration of TERT in almost all cancer cells, the dependency of most cancer cells on TERT activity, and the scarcity of TERT in most normal cells. The approach of transcript knockdown has utilized antisense oligodeoxyribonucleotides as a mainstay, whereas newer advancements have relied on small-interfering RNA (siRNA) molecules. Both these techniques involve synthetic nucleic acids that can bind to specific mRNA targets and both have been effective in anticancer approaches to knockdown TERT expression as described in Chapter 2. The use of double-stranded RNA (dsRNA) has also been quite effective in ablating or greatly reducing transcripts from genes such as *TERT* (*see* Chapter 3). These dsRNA sequences can be used to generate an RNAi response in cells of embryonic origin such as human embryonic kidney (HEK) cells, a popular cell type used in cancer research. This technique is especially effective for short-term analyses of *TERT* knockdown because the dsRNA is degraded in the cells in the long term. RNAi of *TERT* has also been successful with the use of plasmid constructs that exogenously express short hairpin RNA sequences complementary to the *TERT* transcript. This technique (*see* Chapter 4) allows analysis of downstream effects of *TERT*, serves as an alternative approach to gene therapy using viral vectors, and allows long-term and permanent gene knockdown. Also effective for long-term knockdown of *TERT* is the use of retroviral vectors that express short hairpin RNA specific to a segment of the *TERT* transcript. This RNAi-based technique (*see* Chapter 5) involves incorporation of the anti-telomerase sequence into the host genome and can provide effective knockdown of *TERT*.

Small molecules such as 3′-azido-2′, 3′-dideoxythymine (AZT), which is a nucleoside analogue, can be effective in targeting the active site of TERT, but this approach lacks the desired selectivity of many other approaches. Small non-nucleosidic synthetic compounds can be quite effective in inhibiting the catalytic activity of the TERT protein component as described in Chapter 6. One compound that has shown promise in this regard is BIBR1532 that inhibits the in vitro processivity of telomerase. The inhibition of TERT activity with BIBR1532 occurs in a dose-dependent manner, and higher concentrations of this telomerase inhibitor can be cytotoxic to cancer cells of the hematopoietic system such as HL-60 cells with little effect on normal cells.

Anticancer immunotherapeutic approaches have also focused on TERT (*see* Chapter 7). These methods involve the use of peptides derived from TERT.

The peptides are presented by major histocompatibility complex (MHC) class I molecules to T lymphocytes. The result is that CD8+ cytotoxic T lymphocytes specific for the TERT-derived antigenic peptides lyse cancer cells that express TERT. These immunotherapeutic approaches directed against TERT epitopes can be carried out in the absence of toxicity and are showing great promise as anticancer agents.

It can be a challenge to identify small molecule compounds that affect the expression of TERT, and the use of cell-based reporter systems for the analysis of TERT expression has been developed to enhance these endeavors as described in Chapter 8. For example, the *TERT* promoter can be linked to two different reporter genes encoding green fluorescent protein (GFP) and secreted alkaline phosphatase (SEAP). The transfection of these reporter constructs results in stable clones that allow analysis of *TERT* expression. Ultimately, some level of inhibition of *TERT* is the goal of many anticancer approaches, and Chapters 2–8 provide many of the most promising and effective methods for actively knocking down the *TERT* transcript, ablating its catalytic activity, directing the immune system to lyse telomerase-positive cancer cells, or using expression constructs to identify small molecule components that affect the expression of telomerase.

4.2. TR Inhibition as an Anticancer Approach

The RNA component of telomerase has also been a popular and effective target for inhibiting telomerase activity in cancer cells. As in the case for *TERT* transcript knockdown, antisense oligonucleotides against the human TR template can be employed to reduce or eliminate telomerase activity as described in Chapter 9. In this approach, a 2′,5′-oligoadenylate (2–5A) antisense system can be used as a mediator of interferon actions through RNase L activation. The result of this approach is that single-stranded templates, such as the TR component, are specifically cleaved. The anticancer utility of this approach has been proven not only in vitro but also in vivo.

In addition to antisense oligonucleotides, hammerhead ribozymes and RNAi can be directed to the RNA component of telomerase as delineated in Chapter 10. Both these methods also lead to degradation of the RNA component of telomerase. The effect is immediate growth inhibition of cancer cells both in vitro and in vivo independent of telomere length of the target cancer cell. The advantage of this technique is that it greatly reduces the lag period that is often encountered in approaches that are dependent upon the shortening of telomeres to inhibit cancer cell growth. Thus, methods directed at the RNA component of telomerase using antisense oligonucleotides, hammerhead ribozymes, or RNAi also show great promise as anti-telomerase approaches to cancer therapy.

4.3. Targeting Proteins Associated with Telomerase Activity

Approaches to telomerase inhibition have been developed that do not directly inhibit the TERT or TR components of telomerase but rather inhibit target proteins that are associated with telomerase activity. For example, Chapter 11 describes the details of monitoring the telomeric function of tankyrase I, a telomeric poly(ADP-ribose) polymerase (PARP) that can affect telomerase inhibition in cancer cells. The use of Southern blot analysis to screen tankyrase I inhibitors as well as direct monitoring of tankyrase I PARP activity is described.

Signalling pathways such as those carried out by mitogen-activated protein (MAP) kinase can result in stimulation of the *TERT* gene. For example, Ets and AP-1 may play a role in MAP kinase signaling of the *TERT* gene and inhibition of this pathway could be a novel approach to reducing *TERT* expression and telomerase activity as described in Chapter 12. It is apparent that many additional techniques will be developed to impact the proteins or pathways associated with telomerase activity in cancer cells, and Chapters 11 and 12 provide some important approaches for this avenue of potential anticancer therapy.

4.4. Screening of Telomerase Inhibitors

Finding novel inhibitors of telomerase is an important aspect of increasing the tools that we have for anti-telomerase approaches to cancer therapeutics. Chapter 13 proposes a strategy for determining the therapeutic potential of telomerase inhibitors using a screening system in one cell type. For example, four completely different compounds, BRACO19, BIBR1532, 2′-*O*-methyl RNA, and peptide nucleic acids, were chosen for detailing the methods of screening telomerase inhibitors. Additionally, methods are outlined in this chapter for determining the effectiveness of telomerase inhibition through TRAP assays or assessment of telomere lengths using Southern blot telomere restriction fragment analysis.

4.5. Telomerase Inhibition Combined with Other Chemotherapeutic Agents

Chapter 14 reviews the utility of telomerase inhibition in combination with other chemotherapeutic reagents to enhance anticancer effects. For instance, there is an indication that imatinib, a selective inhibitor of the BCR-ABL tyrosine kinase, can enhance the telomerase inhibition of a dominant-negative form of human telomerase (DN-*hTERT*). In a completely different approach, telomestatin, a natural telomerase-inhibiting product isolated from *Streptomyces anulatus*, was combined with imatinib, daunorubicin, mitoxantrone, or vincristine and was shown to enhance the sensitivity of these chemotherapeutic agents. Therefore, approaches to telomerase inhibition may also be merged with completely different anticancer approaches such as chemotherapeutic agents to render more effective modes of cancer therapy.

5. Conclusion

Many of the newest as well as established methods for telomerase-based anticancer approaches are provided in this book. Protocols are presented that involve inhibition of the TERT catalytic subunit of telomerase as well as the TR component of this ribonucleoprotein enzyme. Additional approaches involve intervention directed at the proteins that are associated with telomerase or pathways that modulate the *TERT* gene. Methods for the screening of telomerase inhibitors as well as the potential of merging telomerase inhibition with more conventional chemotherapy are also delineated in this volume. This last concept, that is, combination therapy, may be the most promising approach, and it is likely that many new advances will develop that merge different types of anti-telomerase methods or combine telomerase inhibition with other proven modes of anticancer therapy. The continued development of novel tools will likely be at the forefront of cancer therapy, and this book is intended to provide a synopsis of many different anti-telomerase approaches that may revolutionize cancer therapeutics in the future.

References

1. Ishikawa, F. (1997) Regulation mechanisms of mammalian telomerase. A review. *Biochemistry (Mosc)*. 62, 1332–1337.
2. Weinrich, S.L., Pruzan, R., Ma, L., Ouellette, M., Tesmer, V.M., Holt, S.E., Bodnar, A.G., Lichtsteiner, S., Kim, N.W., Trager, J.B., Taylor, R.D., Carlos, R., Andrews, W.H., Wright, W.E., Shay, J.W., Harley, C.B., Morin, G.B. (1997) Reconstitution of human telomerase with the template RNA component hTR and the catalytic protein subunit hTRT. *Nat. Genet.* 17, 498–502.
3. Beattie, T.L., Zhou, W., Robinson, M.O., Harrington, L. (1998) Reconstitution of human telomerase activity in vitro. *Curr. Biol.* 29, 177–180.
4. Shay, J.W., Bacchetti, S. (1997) A survey of telomerase activity in human cancer. *Eur. J. Cancer* 33, 787–791.
5. Ahmed, A., Tollefsbol, T. (2001) Telomeres and telomerase: basic science implications for aging. *J. Am. Geriatr. Soc.* 49, 1105–1109.

6. Saldanha, S.N., Andrews, L.G., Tollefsbol, T.O. (2003) Analysis of telomerase activity and detection of its catalytic subunit, hTERT. *Anal. Biochem.* 315, 1–21.
7. Hodes, R. (2001) Molecular targeting of cancer: telomeres as targets. *Proc. Natl. Acad. Sci. U. S. A.* 98, 7649–7651.
8. Ahmed, A., Tollefsbol, T. (2003) Telomeres, telomerase, and telomerase inhibition: clinical implications for cancer. *J. Am. Geriatr. Soc.* 51, 116–122.

2

Telomerase Inhibition by Synthetic Nucleic Acids and Chemosensitization in Human Bladder Cancer Cell Lines

Kai Kraemer, Susanne Fuessel, and Axel Meye

Summary

The knockdown of genes that are over-expressed in cancer, and function in tumor onset and/or progression, is an attractive tool to impair the growth of tumor cells. Synthetic nucleic acids such as antisense oligodeoxynucleotides (AS-ODNs) or small-interfering RNAs (siRNAs) were applied against different tumor-associated transcripts, including the human telomerase reverse transcriptase (hTERT), to inhibit the proliferation of tumor cells and to sensitize them against chemotherapeutic (CT) agents. The efficacy of nucleic acid-based inhibitors was evaluated in vitro by determining the extent of down-regulation of the respective target mRNA and protein expression as well as by extensively investigating growth properties (e.g., viability, proliferation, apoptosis, and cell-cycle distribution) of the affected tumor cells. Methods for a successful down-regulation of hTERT and for the quantitative determination of resulting effects on cellular growth were described herein.

Key Words: Antisense oligodeoxynucleotides; apoptosis; bladder cancer; chemotherapeutics; human telomerase reverse transcriptase; proliferation; small-interfering RNAs; transfection; viability.

1. Introduction

The human telomerase reverse transcriptase (hTERT), the catalytically active component of the telomerase complex (1), catalyzes the telomere elongation and associates with telomeres leading to increased genomic stability. The hTERT-mediated enhanced DNA repair (2) could contribute to resistance of

From: *Methods in Molecular Biology, vol. 405: Telomerase Inhibition*
Edited by: L. G. Andrews and T. O. Tollefsbol © Humana Press Inc., Totowa, NJ

tumor cells against genotoxic drugs. Thus, telomerase inhibition may cause an increase in sensitivity to chemotherapeutics (CTs) in tumor cells (*3,4*). Hence, the close association of hTERT with the tumorigenic process supports the usage of hTERT as an useful and specific anti-tumor target.

The treatment of cancer cells with nucleic acid-based inhibitors provides a suitable approach to affect tumor cells, specifically by suppressing growth-promoting genes. Inhibitors such as antisense oligodeoxynucleotides (AS-ODNs) and small-interfering RNAs (siRNAs) are short synthetic nucleic acids that are able to bind a selected target mRNA in a sequence specific manner by Watson–Crick base pairing. Single-stranded AS-ODNs can interfere with translation and/or trigger the RNase H-mediated cleavage of the target mRNA (*5*). The action of siRNAs—short double-stranded RNA molecules with characteristic 3′ overhangs—in mammalian cells is based on the formation of the RNA-induced silencing complex (RISC). The nuclease activity of RISC cleaves the target mRNA after sequence-specific hybridization with the antisense strand of the siRNA molecule (*6*). Whether AS-ODNs or siRNAs are more efficient or more specific is controversially discussed (*7–9*).

The design of nucleic acid-based inhibitors that function by complementary base pairing is important for their successful application. The identification of putatively accessible single-stranded regions within the target mRNA by secondary structure prediction may be a helpful tool to find efficient AS-ODNs (*10*) and siRNAs (*11*).

The prerequisite of a potent gene knockdown is the efficient uptake of nucleic acid-based inhibitors within cells. A widely used in vitro method is the transfection of AS-ODNs or siRNAs, using cationic lipid formulations or other transfection reagents such as dendrimers, which transfer the constructs through the cell membrane (*12*).

Our previous studies were concentrated on the investigation of AS-ODNs directed at the hTERT mRNA and their effects on human bladder cancer (BCa) cells in vitro (*13,14*). The carcinoma of the bladder is the second most common malignancy in the urinary tract. hTERT was shown to be highly specific expressed in BCa cells but not detectable in normal urothelial cells (*15*). The treatment of BCa cells in vitro with hTERT-directed AS-ODNs caused massive effects on the growth of tumor cells (*13*). Furthermore, these AS-ODNs (ASt2206 and ASt2331) enhanced the growth inhibitory action of CTs in several BCa cell lines (*14*). In these previous studies, commercially available kits, which are described herein, were used to characterize the cellular growth at different levels (viability, proliferation, colony-formation assay, apoptosis, and cell-cycle distribution).

2. Materials

2.1. Design of AS-ODNs and siRNAs

1. Software Mfold for calculation of the secondary structure of the hTERT mRNA (http://www.bioinfo.rpi.edu/~zukerm/rna/).
2. siRNA design tool (http://www.qiagen.com).
3. BLASTn database search (http://www.ncbi.nlm.nih.gov/blast/).

2.2. Cell Culture, Transfection Reagents, and CTs

1. Dulbecco's modified Eagle's medium (DMEM; Invitrogen, Karlsruhe, Germany) supplemented with 10% fetal calf serum (FCS; Invitrogen), 1% *N*-2-hydroxyethylpiperazine-*N'*-2-ethane sulfonic acid (HEPES) buffer (Invitrogen), and 1% minimum essential medium (MEM) non-essential amino acids (Invitrogen). RPMI 1640 medium supplemented with 10% FCS (Invitrogen). MEM with 10% FCS (Invitrogen).
2. Solution of trypsin (0.25%) and ethylenediamine tetraacetic acid (EDTA; 1 mM) (Invitrogen).
3. Serum-free culture medium OPTIMEM 1 (Invitrogen).
4. ODNs (20 mer, desalted) containing 2 phosphorothioates at the 5′ and 3′ ends, respectively (Invitrogen); High-Performance Purity Grade siRNAs (Qiagen, Hilden Germany). ODN sequences were selected using design algorithms mentioned in **Subheading 2.1**.
5. *N*-[1-(2,3-Dioleoyloxy)propyl]-*N*,*N*,*N*-trimethyl-ammonium-methysulfate (DOTAP) Liposomal Transfection Reagent (Roche, Mannheim, Germany); Lipofectin (LF) Transfection Reagent (Invitrogen).
6. Mitomycin C (medac, Wedel, Germany, or Sigma Aldrich, St. Louis, MO), gemcitabine (GEM; Eli Lilly, Indianapolis, IN), and cisplatin (CDDP; Gry-Pharma, Kirchzarten, Germany, or Sigma Aldrich) freshly diluted in culture medium before use.
7. Cell and Particle Counter Coulter Z2 (Beckman-Coulter, Krefeld, Germany) or any other cell counter or counting chamber.

2.3. Cellular Uptake of Nucleic Acid-Based Inhibitors

1. Fluorescence-labeled ODNs containing a 5′-carboxyfluorescein (FAM) modification, siRNAs labeled at the 3′ end of the sense strand with Alexa Fluor 488 (both from Invitrogen, Qiagen, (Eurogentec, Seraing, Belgium); (MWG Biotech, Ebersberg, Germany); or other providers).
2. 4-Well culture slides (BD Biosciences, Heidelberg, Germany); Fluoromount-G mounting medium (SouthernBiotech, Birmingham, AL).
3. 4% Buffered formalin for fixation of cells.

4. FACScan flow cytometer (BD Biosciences); Axioskop 2 MOT fluorescence microscope (Carl Zeiss, Goettingen, Germany).
5. WinMDI 2.8 (http://facs.scripps.edu/software.html) free software for analyzing data from fluorescence-activated cell sorting (FACS).

2.4. Investigations of the Effects on Target Expression and Cell Growth

1. RNA isolation: Invisorb Spin Cell RNA Mini Kit (Invitek, Berlin, Germany) or comparable RNA isolation kits from other providers.
2. hTERT expression: LightCycler Telo*TAGGG*hTERT Quantification Kit (cat. no. 03 012 344 001, Roche) or self-designed or premade gene expression assays from different providers (e.g., TaqMan assays from Applied Biosystems).
3. LightCycler instrument (Roche); LightCycler Software 3.5 (Roche); other real-time PCR devices.
4. Viability assay: Cell-Proliferation Reagent water soluble tetrazolium-1 (WST-1, Roche) or comparable assays for the determination of metabolic activity (e.g., MTT assay from Roche).
5. Colony-formation assay: 4% buffered formalin; Giemsa solution (Merck, Darmstadt, Germany; freshly filtered, 1:20 in aqua bidest).
6. Apoptosis assay: Annexin V FITC Apoptosis Detection Kit I (BD Biosciences).
7. Cell-cycle analysis: CycleTest Plus DNA Reagent Kit (BD Biosciences); Software ModFit LT 3.0 (Verity Software House, Topsham, ME).

3. Methods
3.1. Design of AS-ODNs and siRNAs

1. To increase the probability for the identification of effective AS-ODNs, the secondary structure of the hTERT mRNA (Accession number AF015950) was calculated using the Software Mfold 2.3 (*see* **Note 1**). Owing to the length of the mRNA of 4015 nucleotides, the whole sequence was divided into multiple overlapping windows. For each window, the 10 putatively most stable structures with the lowest free energy were predicted and scanned for conserved single-stranded sequence motifs with at least nine nucleotides in length (*see* **Note 2**). Against each of five selected motifs, four AS-ODNs were designed. All AS-ODNs were checked for non-target binding capacities by BLASTn database search. Sequences of the most potent anti-hTERT AS-ODNs are given in **Table 1**.
2. The hTERT-directed siRNAs were designed and synthesized by Qiagen using a siRNA design algorithm with neural network technology that is based on a large database of experimental data (*see* **Note 3**) and **Table 1**).
3. For both AS-ODNs and siRNAs, suitable controls without homology to human mRNAs are necessary for normalization of the results (*see* **Note 3** and **Table 1**).

Table 1
Construct sequences of Antisense Oligodeoxynucleotides (AS-ODNs) and mRNA Target Sequences of Small-Interfering RNAs (siRNAs) Directed at Human Telomerase Reverse Transcriptase (hTERT) (AF015950) and of Control Constructs

Sequences	
ODN (5′–3′)	
ASt2206 *(13)*	TGTCCTGGGGGATGGTGTCG
ASt2315 *(13)*	TTGAAGGCCTTGCGGACGTG
ASt2317 *(13)*	TCTTGAAGGCCTTGCGGACG
ASt2331 *(13)*	GGTAGAGACGTGGCTCTTGA
ASt2333 *(13)*	AAGGTAGAGACGTGGCTCTT
siRNAs (5′–3′)	
si-hTERT1	CUGGAGCAAGUUGCAAAGCAU
si-hTERT2	CAGCUCCCAUUUCAUCAGCAA
Control constructs (5′–3′)	
NS-K1 *(17)*	TAAGCTGTTCTATGTGTT
NS-si [RNAi Human/Mouse Control Kit (Qiagen)]	AAUUCUCCGAACGUGUCACGU
si-FITC (MWG Biotech)	CGUACGCGGAAUACUUCGA

3.2. Cell Cultivation, Transfection Experiments, and Chemotherapy

1. All human BCa cell lines were cultivated under standard conditions in antibiotic-free medium (EJ28—DMEM; 5637 and RT112—RPMI 1640, J82—MEM).
2. The cells were seeded in culture plates and grown for 3 days to reach a growth density of 50–70%. Selection of appropriate culture plates depended on the planned analyses (96-well for viability assays and 6-well for isolation of RNA or protein and for analyses of apoptosis or cell cycle). Numbers of seeded cells were chosen according to previous optimization experiments and depended on the doubling time of the selected cell lines and the planned period of analyses (have to be optimized for each of these parameters). Suggested cell numbers for seeding of BCa cells and incubation for 72 h until treatment are given in **Table 2**.

   For transfection, a mixture of ODNs and OPTIMEM 1 as well as a mixture of LF and OPTIMEM 1 were prepared. After an incubation period of 15 min at room temperature, both mixtures were combined, mixed thoroughly, and incubated again for 30 min. We used a final ODN concentration of 250 nM and a LF : ODN ratio of 3:1 (w/w). The cells were transfected with the ODN/LF mixture for 4 h under culture

Table 2
Cell numbers for seeding BCa cells in different culture 72 h before treatment

Culture vessel	Area/well	Volume/well	Cell number
6-well plate	$9.6\,cm^2$	2 ml	$3–8 \times 10^4$
24-well plate/4-well slide	$1.9/1.7\,cm^2$	$600\,\mu l$	$1–5 \times 10^4$
96-well plate	$0.35\,cm^2$	$100\,\mu l$	$1–4 \times 10^3$

conditions followed by washing with phosphate-buffered saline (PBS) and incubation in fresh culture medium until further use. For the siRNA transfection with LF, an identical procedure was applied (*see* **Note 5**).

3. The DOTAP/nucleic acid mixture in OPTIMEM 1 can be used after thorough shaking without further incubation. For the cell lines used, a DOTAP : AS-ODN ratio of 3:1 (w/w) and a DOTAP : siRNA ratio of 4:1 (w/w) were most efficient in mediating the cellular uptake at tolerable cellular toxicity (*see* **Note 6**).

4. To combine an hTERT inhibition with CTs, pretreatment with AS-ODNs or siRNAs was performed as stated in **Subheading 1**, **paragraphs 2** or **3**, followed by incubation with CT after 24 h. The cells were incubated with mitomycin C (MMC) (2 h), GEM (24 h), or CDDP (24 h). The CTs were dissolved in PBS and diluted in culture medium to the appropriate concentrations (*see* **Note 7**). The cellular effects were investigated 48 and 72 h after start of the treatment with CT (*see* **Subheading 3.4.**) *(14)*.

3.3. Cellular Uptake of Nucleic Acid-Based Inhibitors

1. The cells were seeded in 4-well culture slides for microscopic studies or 6-well culture plates for FACS analyses (*see* **Table 2**). After 3 days, cells were transfected as stated in **Subheading 3.2**, **steps 2** and **3**, using different ratios [between 3:1 and 4:1 (w/w)] of the respective transfection reagent and of the fluorescent-labeled ODNs or siRNAs.

2. Fluorescence microscopy was performed at different time points after transfection (e.g., directly, 24, 48, and 72 h). The medium was removed from the chambers, the cells were washed with PBS ($500\,\mu l$/chamber), fixed with 4% formalin for 20 min, and washed again twice by PBS. Afterward, the slides were mounted using $50–80\,\mu l$ Fluoromount-G mounting medium followed by sealing of the coverslips with nail polish (which should not get under the coverslips) and observation under the fluorescence microscope.

3. For FACS analysis, the cells were harvested by mild digestion with 0.05% trypsin/0.02% EDTA (at least 5 min at 37 °C) and washed with PBS. To discriminate

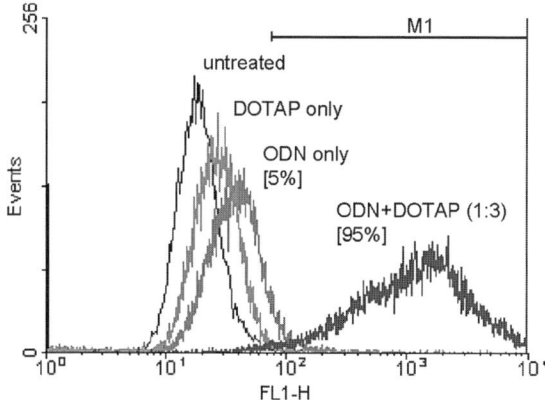

Fig. 1. The histogram shows the increase in fluorescence intensity of 5637 bladder cancer (BCa) cells after transfection with 5′-carboxyfluorescein (FAM)-labeled oligodeoxynucleotides (ODNs) (FL1-H channel). The marker M1 was set to quantify the cells showing increased fluorescence intensity in comparison with cells treated by *N*-[1-(2,3-dioleoyloxy)propyl]-*N,N,N*-trimethyl-ammonium-methysulfate (DOTAP) only. The DOTAP-mediated transfection caused a fraction of FAM-positive cells of 95%, whereas only 5% of the cells incorporated the ODNs without complexation with DOTAP.

between viable and dead cells, a counterstain with $0.5\,\mu g/ml$ propidium iodide was performed before measuring 2×10^4 cells/sample.

4. The Software WinMDI 2.8 was applied for data analysis. A cell population with normal size and granularity (discriminated by forward-scatter and side-scatter detection) was investigated. The increase in FITC fluorescence intensity after treatment with labeled ODNs or siRNAs was assessed by comparison with untreated cells and cells treated by the transfection reagent only (*see* **Fig. 1**).

3.4. Investigation of the Effects on Target Expression and Cell Growth

3.4.1. Analysis of the hTERT mRNA Expression

1. A fast and efficient spin column-based method to isolate total RNA from cell cultures is the Invisorb Spin Cell RNA Mini Kit. It allows the RNA isolation from 24 samples in less than 1 h (*see* **Note 8**).
2. The hTERT mRNA can be quantified using the LightCycler Telo*TAGGG*hTERT Quantification Kit. The procedure includes the reverse transcription of the samples followed by quantification of hTERT and the reference gene porphobilinogen deaminase (*PBGD*) in the same PCR run (*see* **Note 9**).

3.4.2. Examination of Cell Growth

The WST-1 assay exhibits a suitable standard procedure for high-throughput screening of changes in cellular growth. A reduced viability may be caused by increased apoptosis and/or cell-cycle arrest. Therefore, quantification of apoptosis and determination of cell-cycle alterations are important for the detailed characterization of growth inhibitory effects.

1. The WST-1 assay is performed at different time points (24, 48, 72, and 96 h) after transfection in 96-well culture plates (*see* **Table 2**). The transfected cells–incubated at least in triplicates – were supplemented with 10 µl WST-1, which was added directly into the culture medium, followed by incubation for 1–4 h. The viability was measured by determining the optical density using a spectrophotometer for 96-well plates at 450 nm against a background of 620 nm. WST-1 is non-toxic, thus the treatment can be repeated after washing the cells with PBS and further incubation in culture medium (*see* **Note 10**).

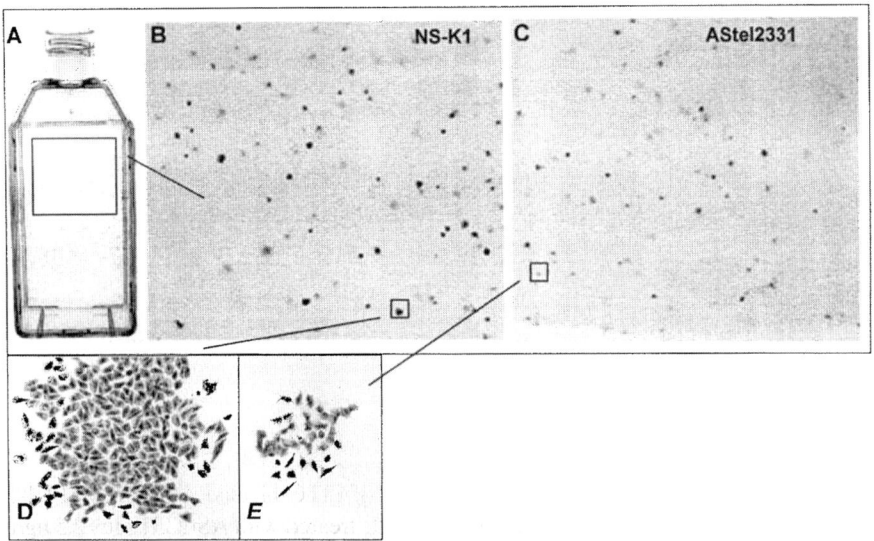

Fig. 2. Examples of clonogenic survival of EJ28 cells treated with anti-human telomerase reverse transcriptase (hTERT) antisense oligodeoxynucleotide (AS-ODN) or control ODN (after Giemsa staining). Macroscopic (**A**—25-cm² culture flask; **B** and **C**—enlarged sections) and microscopic (**D** and **E**—single colonies at larger magnification) pictures of cells transfected with AStel2331 (**C** and **E**) or NS-K1 (**B** and **D**), respectively.

2. The ability of single cells to adhere and to form colonies was assessed by the colony-formation assay. The cells were harvested at defined periods after transfection (after 24, 48, or 72 h), highly diluted in culture medium and seeded at least in triplicates in 6-well plates (*see* **Note 11**). After a suitable time (normally 7–12 days), the cells were fixed in 4% formalin for 20 min, washed with PBS, and stained with Giemsa solution (1:20) for 10 min. After two washings with aqua bidest and drying at the air, the formed colonies were counted under the light microscope (*see* **Note 12** and **Fig. 2**).

3. For quantification of apoptosis by FACS, the Annexin V FITC Apoptosis Detection Kit I is a simple and robust procedure and was performed according to kit instructions. The cells (1×10^5–1×10^6) were harvested by trypsin/EDTA treatment (*see* **Note 13**), washed with PBS, and resuspended in 100 µl Annexin V binding buffer. After addition of 5 µl Annexin V FITC and 5 µl propidium iodide (PI), the cells were incubated for 15 min in the dark followed by addition of 400 µl Annexin V binding buffer and FACS measurement of 2×10^4 cells/sample. Data were examined using the quadrant analysis mode of WinMDI 2.8, which allows the discrimination of viable cells, early apoptosis, late apoptosis, and necrosis (*see* **Fig. 3**).

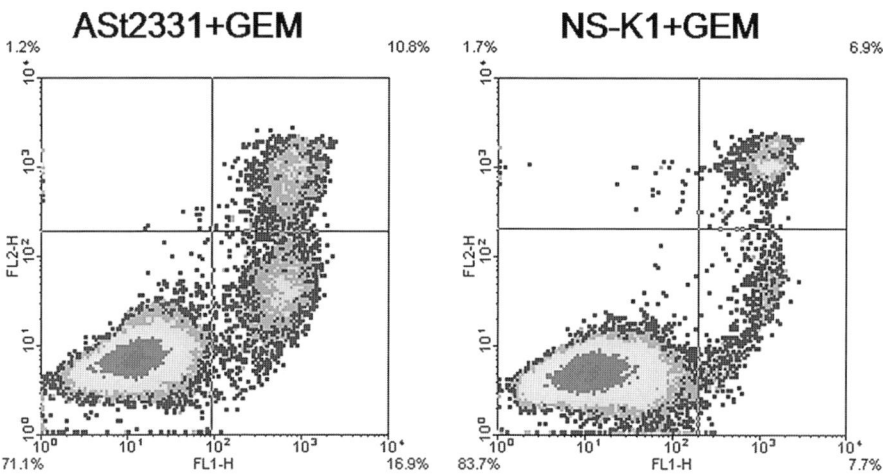

Fig. 3. Plotted are the fluorescence intensities of FITC-labeled Annexin V (FL1-H) versus propidium iodide (PI) (FL2-H) for EJ28 cells treated with ASt2331 plus 2.5 ng/ml gemcitabine (GEM) or with a non-sense ODN (NS-K1) plus GEM, respectively. After 4 h of ODN transfection, incubation in culture medium for 20 h, and subsequent treatment with 2.5 ng/ml GEM for 24 h, quantification of apoptosis was performed 72 h after start of the transfection as described *(14)*. The quadrant analysis (relative percentages of cells in the quadrants are given at the corners of each plot) allows classification and quantification of viable cells (lower left), cells undergoing early apoptosis (lower right), necrotic cells (upper left), and cells died by apoptosis (upper right).

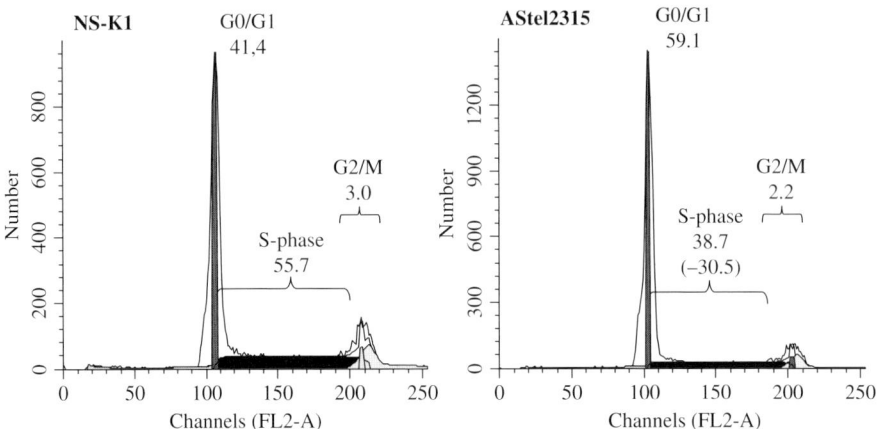

Fig. 4. Cell-cycle alterations after oligodeoxynucleotide (ODN) treatment of EJ28 cells. Cells treated with the human telomerase reverse transcriptase (hTERT)-directed antisense ODN (AS-ODN) AStel2315 or the control ODN NS-K1 were harvested 24 h after transfection. After propidium iodide (PI) staining of DNA using the CycleTest Plus DNA Reagent Kit analyses of cell-cycle distribution were performed by flow cytometry (channel FL2). Percentage of each cell-cycle phase (G0/G1, S, and G2/M) was assessed by the Software ModFit LT 3.0 and compared between the different treatments.

4. The CycleTest Plus DNA Reagent Kit enables the quantification of cell populations within the phases of the cell cycle (*see* **Fig. 4**). The harvested and washed
   cells were resuspended in 1 ml buffer solution (supplied with the kit) and immediately stained or frozen at $-80\,°C$. After removing the buffer solution, cell
   membranes and cytoskeletons were digested by incubating the cells in $250\,\mu l$
   trypsin containing solution A for 10 min. The addition of $200\,\mu l$ solution B
   affects the inhibition of trypsin activity and the digestion of RNA. The samples
   were stained by incubation in $200\,\mu l$ solution C (PI solution) for 10 min in
   the dark at $2–8\,°C$ followed by FACS measurement within 3 h after addition
   of solution C.

4. Notes

1. Owing to the possibility of calculating RNA sequences of up to 6000 nucleotides
   using the latest available Mfold version 3.1, splitting of the mRNA into single
   windows is not necessary. In addition to Mfold, different kinds of software
   (RNAstructure, RNAdraw, Vienna RNA Package, etc.) are available for secondary
   structure calculation. An overview is given on the RNA world homepage of the
   IMB Jena (http://www.imb-jena.de/RNA.html#Software).

2. In our experience, the potency of the AS-ODNs investigated depended on an appropriate single-stranded sequence motif but seemed to be dependent on neither the length nor the type (loops, joints, and bulges) of the motif.

3. A multitude of siRNA design tools is available in the world wide web, for example, the Qiagen design algorithm (http://www.qiagen.com), siDirect (http://alps3.gi.k.u-tokyo.ac.jp/~yamada/sdirect2/index.php), and siRNA Target Finder (https://www.genscript.com/ssl-bin/app/rnai). Furthermore, several providers offer the design of custom siRNAs or premade siRNAs (e.g., Invitrogen, and Applied Biosystems).

4. We recommend the comparative examination of additional controls for proper assessment of construct specificity: (i) already established AS-ODNs or siRNAs directed at other targets and (ii) mismatch controls containing 1–4 base exchanges in comparison with the primary construct.

5. A working solution at $50\,\mu M$ was prepared in PBS from all ODNs. The siRNAs were diluted in siRNA suspension buffer (supplied by Qiagen) at a concentration of $20\,\mu M$. All solutions were prepared under sterile conditions.

6. The optimization of the transfection conditions should be adapted to each cell type, regarding to efficient uptake and low toxicity. According to our experience, AS-ODNs and siRNAs were efficiently transfected into different BCa cell lines (EJ28, 5637, RT112, and J82) using LF and DOTAP, whereby the toxicity of LF was generally higher. Nevertheless, any other transfection reagent can be used, but transfection conditions, which depend on the cell line and type of constructs, should be optimized.

7. Each drug concentration was adapted to the appropriate cell line to reach a moderate inhibition of viability. Multiple CT doses [generally the inhibitory concentration of 50% (IC_{50}) and the IC_{70}] should be used for the combination experiments. For a proper interpretation of the results on cellular growth obtained by combination of AS-ODNs or siRNAs and CT, the following controls are recommended: (i) untreated cells, (ii) cells treated by CT only, (iii) cells treated by AS-ODNs or siRNAs only, and (iv) cells treated by control constructs in combination with CT (*see* **Fig. 5**).

8. Standard values for the RNA yield by the Invisorb Spin Cell RNA Mini Kit are $4–5\,\mu g$ total RNA from 2×10^5 BCa cells and $>1\,\mu g$ from 5×10^4 BCa cells. The volume of elution buffer needed to remove the RNA from the column should be adapted to the initial cell number per sample to increase the concentration of low amounts of RNA (for cell numbers $<1 \times 10^5$, $20–30\,\mu l$ elution buffer are recommended).

9. The LightCycler Telo*TAGGG*hTERT Quantification Kit is relatively expensive but has the advantage of including the reverse transcription of the samples. Furthermore, one can get a complete set of primer and hybridization probes not only for hTERT but also for the reference gene *PBGD*. Thus, the fast quantification (1 h) of 12 samples per RT-PCR run is possible. We recommended the

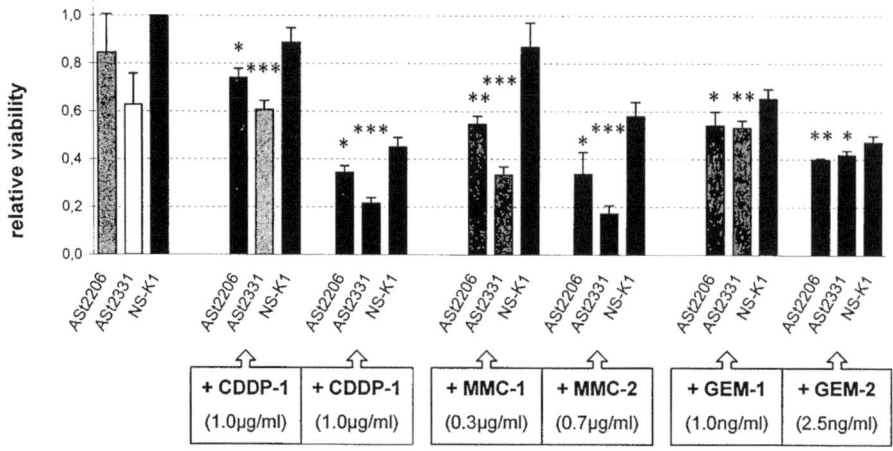

Fig. 5. Enhancement of the chemotherapeutic (CT)-mediated effects on viability by pretreatment with two different human telomerase reverse transcriptase (hTERT) antisense oligodeoxynucleotides (AS-ODNs) (ASt2206 and ASt2331) in EJ28 cells 96 h after transfection. The viability of cells treated with non-sense ODN (NS-K1) represents 100% (1.0). Two doses were used for each CT [cisplatin (CDDP), mitomycin C (MMC), and gemcitabine (GEM)]. Error bars represent standard deviation of quadruplicates. Asterisks indicate significant differences between AS-ODN plus CT and NS-K1 plus CT calculated by Student's t-test ($^*p < 0.05$, $^{**}p < 0.01$, and $^{***}p < 0.001$), modified from **ref. (14)**.

application of 100 ng total RNA per PCR. Alternatively, an in-house assay for the quantification of the transcript amounts hTERT and an eligible reference gene using a real-time PCR system can be established and applied *(15,16)*.

10. The WST-1 incubation time depends on the cell line and should be previously optimized to ensure the measurement within a range where the optical density depends on the cell number in a linear manner. We recommend the regular observation of treated cells by light microscopy to confirm the results obtained by WST-1 and to detect outliers (e.g., through incorrect seeding or absent WST-1 reagent). A reduction in viability should be always seen by microscopic observation. Other assays for determination of metabolic activity might serve as alternative tools. However, these assays frequently necessitate a fixation of cells and thereby do not allow repeated measurements. Furthermore, trypan blue exclusion assay is a commonly used method for evaluation of cell growth but does not permit a valid quantification such as user-independent measurements of absorbance changes.

11. The dilution to a single-cell level is a critical step for the colony-formation assay. Careful and fast pipetting is necessary. The optimal cell number for seeding of

each cell line should be determined. It depends on the proliferation rate and lies in a range of 100–500 cells/well for the BCa cell lines tested in our laboratory.

12. The time to terminate the colony growth should be determined by light microscopy. The colonies should not overlap, but they should reach a proper size of more than 20 cells/colony. Alternatively to microscopic counting, all macroscopically visible colonies can be counted.

13. Apoptotic cells may detach from the culture plate and float within the medium. To consider these cells for apoptosis detection, the culture supernatant should be collected and pooled with the monolayer cells harvested by trypsin/EDTA.

Acknowledgments

We thank all colleagues from our department for their support as well as Dr M. Kotzsch (Institute for Pathology) and Dr B. Schwenzer (Department of Biochemistry) for permanent encouragement. Furthermore, we express our gratitude to the Pinguin foundation for its grant to K.K. and the Robert Pfleger foundation for its grant to A.M. and S.F.

References

1. Nakamura, T. M., Morin, G. B., Chapman, K. B., Weinrich, S. L., Andrews, W. H., Lingner, J., Harley, C. B., and Cech, T. R. (1997) Telomerase catalytic subunit homologs from fission yeast and human. *Science* **277**, 955–9.

2. Sharma, G. G., Gupta, A., Wang, H., Scherthan, H., Dhar, S., Gandhi, V., Iliakis, G., Shay, J. W., Young, C. S., and Pandita, T. K. (2003) hTERT associates with human telomeres and enhances genomic stability and DNA repair. *Oncogene* **22**, 131–46.

3. Chen, Z., Koeneman, K. S., and Corey, D. R. (2003) Consequences of telomerase inhibition and combination treatments for the proliferation of cancer cells. *Cancer Res* **63**, 5917–25.

4. Misawa, M., Tauchi, T., Sashida, G., Nakajima, A., Abe, K., Ohyashiki, J. H., and Ohyashiki, K. (2002) Inhibition of human telomerase enhances the effect of chemotherapeutic agents in lung cancer cells. *Int J Oncol* **21**, 1087–92.

5. Crooke, S. T. (1999) Molecular mechanisms of action of antisense drugs. *Biochim Biophys Acta* **1489**, 31–44.

6. Agrawal, N., Dasaradhi, P. V., Mohmmed, A., Malhotra, P., Bhatnagar, R. K., and Mukherjee, S. K. (2003) RNA interference: biology, mechanism, and applications. *Microbiol Mol Biol Rev* **67**, 657–85.

7. Bilanges, B., and Stokoe, D. (2005) Direct comparison of the specificity of gene silencing using antisense oligonucleotides and RNAi. *Biochem J* **388**, 573–83.

8. Grunweller, A., Wyszko, E., Bieber, B., Jahnel, R., Erdmann, V. A., and Kurreck, J. (2003) Comparison of different antisense strategies in mammalian cells using locked nucleic acids, 2′-O-methyl RNA, phosphorothioates and small interfering RNA. *Nucleic Acids Res* **31**, 3185–93.

9. Tsui, P., Rubenstein, M., and Guinan, P. (2005) siRNA is not more effective than a first generation antisense oligonucleotide when directed against EGFR in the treatment of PC-3 prostate cancer. *In Vivo* **19**, 653–6.

10. Patzel, V., Steidl, U., Kronenwett, R., Haas, R., and Sczakiel, G. (1999) A theoretical approach to select effective antisense oligodeoxyribonucleotides at high statistical probability. *Nucleic Acids Res* **27**, 4328–34.

11. Kretschmer-Kazemi Far, R., and Sczakiel, G. (2003) The activity of siRNA in mammalian cells is related to structural target accessibility: a comparison with antisense oligonucleotides. *Nucleic Acids Res* **31**, 4417–24.

12. Axel, D. I., Spyridopoulos, I., Riessen, R., Runge, H., Viebahn, R., and Karsch, K. R. (2000) Toxicity, uptake kinetics and efficacy of new transfection reagents: increase of oligonucleotide uptake. *J Vasc Res* **37**, 221–34; discussion 303–4.

13. Kraemer, K., Fuessel, S., Schmidt, U., Kotzsch, M., Schwenzer, B., Wirth, M. P., and Meye, A. (2003) Antisense-mediated hTERT inhibition specifically reduces the growth of human bladder cancer cells. *Clin Cancer Res* **9**, 3794–800.

14. Kraemer, K., Fuessel, S., Kotzsch, M., Ning, S., Schmidt, U., Wirth, M. P., and Meye, A. (2004) Chemosensitization of bladder cancer cell lines by human telomerase reverse transcriptase antisense treatment. *J Urol* **172**, 2023–8.

15. De Kok, J. B., Schalken, J. A., Aalders, T. W., Ruers, T. J., Willems, H. L., and Swinkels, D. W. (2000) Quantitative measurement of telomerase reverse transcriptase (hTERT) mRNA in urothelial cell carcinomas. *Int J Cancer* **87**, 217–20.

16. Yajima, T., Yagihashi, A., Kameshima, H., Furuya, D., Kobayashi, D., Hirata, K., and Watanabe, N. (2000) Establishment of quantitative reverse transcription–polymerase chain reaction assays for human telomerase-associated genes. *Clin Chim Acta* **290**, 117–27.

17. Chen, J., Wu, W., Tahir, S. K., Kroeger, P. E., Rosenberg, S. H., Cowsert, L. M., Bennett, F., Krajewski, S., Krajewska, M., Welsh, K., Reed, J. C., and Ng, S. C. (2000) Down-regulation of survivin by antisense oligonucleotides increases apoptosis, inhibits cytokinesis and anchorage-independent growth. *Neoplasia* **2**, 235–41.

3

hTERT Knockdown in Human Embryonic Kidney Cells Using Double-Stranded RNA

Serene R. Lai, Lucy G. Andrews, and Trygve O. Tollefsbol

Summary

The method of RNA interference (RNAi) is an easy means of knocking down a gene without having to generate knockout mutants, which may prove to be difficult and time consuming. RNAi is a naturally occurring process that involves targeting the mRNA of a gene by introducing RNAs that are complementary to the target mRNA. The foreign RNAs activate an endogenous enzyme, DICER, which degrades the target mRNA. There are many ways of eliciting the RNAi response in a cell. In this chapter, we describe the use of double-stranded RNA (dsRNA) to knockdown human telomerase reverse transcriptase (*hTERT*), the gene that codes for the catalytic subunit of the human telomerase enzyme. dsRNA can be used to generate the RNAi response in cells of embryonic origin, such as human embryonic kidney (HEK) cells. The RNAi effect is transient because the dsRNA eventually gets degraded in the cells, and it is useful to study the short-term effects of a gene knockdown.

Key Words: dsRNA; RNAi; HEK cells; hTERT.

1. Introduction

In most mammalian cells, the introduction of a long double-stranded RNA (dsRNA) of more than 30 bp induces an interferon response that activates RNase L, which causes a non-specific systemic degradation of mRNA in the cell. This results in the shutdown of protein translation and further leads to apoptosis *(1)*. It has been suggested that this interferon response evolved to inhibit viral gene expression in infected cells *(2)*. However, this interferon response appears to be absent in mammalian cells of embryonic origin *(3)*.

From: *Methods in Molecular Biology, vol. 405: Telomerase Inhibition*
Edited by: L. G. Andrews and T. O. Tollefsbol © Humana Press Inc., Totowa, NJ

After long dsRNA has been introduced into mammalian embryonic cells, DICER, an RNase, cleaves the dsRNA into 21–23 nucleotide-long fragments of short-interfering RNAs (siRNAs) *(4)*. These siRNAs are further incorporated into an RNA-induced silencing complex (RISC) where the siRNAs are unwound into single-stranded RNA (ssRNA). The RISC is involved in recruiting the ssRNA to the mRNA of complementary sequence and cleaving the mRNA, thus preventing translation *(5,6)*. This allows for target-specific knockdown of a gene product. The advantage of using dsRNA for knocking down genes is that multiple siRNA targets are generated to the specific mRNA, making this a more efficient method of RNA interference (RNAi) as compared with using a single siRNA target. However, this may also result in an increase in non-specific knockdown of other genes because of sequence homology *(7,8)*.

It has been well-documented that the efficacy of gene target knockdown is variable depending on the length of dsRNA and selection of target sequence. Preferably a length of more than 200 bp is used for dsRNAi for an optimal knockdown effect. Also, the target sequence should contain only 30–50% G-C content *(9)*. This chapter describes the methods for human telomerase reverse transcriptase (*hTERT*) gene knockdown using dsRNA in human embryonic kidney (HEK) cells. The use of dsRNA allows for a transient knockdown of *hTERT* to determine its immediate effects, because the DICER-generated siRNAs will eventually degrade in the cell. Cellular levels of hTERT mRNA have been shown to be maximally depleted in the cell from 24 to 72 h, and its levels were almost recovered by day 6 of transfection (*see* **Fig. 1**).

hTERT is the catalytic subunit of human telomerase, the enzyme responsible for maintaining telomere length, contributing to chromosomal stability. *hTERT* knockdown potentially has many applications in cancer therapy, because telomerase is necessary for the unlimited division of 90–95% of human cancer cells *(10)*. Although telomerase is up-regulated in these cancer cells, most cancer cells have shorter telomeres than normal cells *(11)*. The use of a transient and reversible method of knocking down hTERT, such as the use of dsRNAi, is thus highly beneficial because the regulation of telomerase activity can be performed and discontinued before telomeres of normal cells reach critical length. The recovery from dsRNA treatment is expedient, because the recovery of telomerase levels is almost complete by day 6 of treatment with dsRNA (*see* **Fig. 1**), thus it is easy to initiate and terminate the knockdown of hTERT. Telomeres of germ and stem cells are also more likely to regain their telomere lengths after therapy with dsRNA *(12)*.

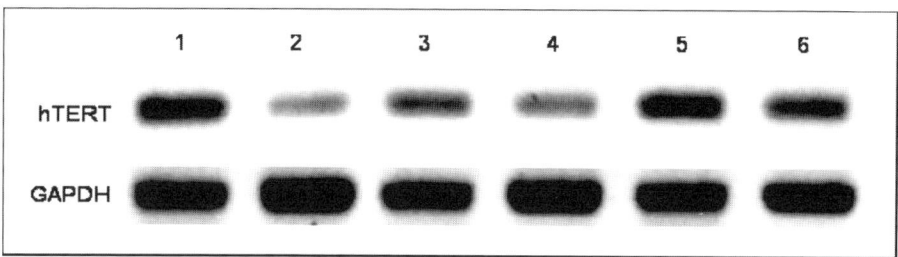

Fig. 1. Agarose gel electrophoresis from reverse-transcriptase PCRs using human telomerase reverse transcriptase (hTERT) and GAPDH (glyceraldehyde phosphate dehydrogenase) primers. hTERT knockdown in human embryonic kidney (HEK) cells was performed by transfecting with 2.5 μg 219 bp double-stranded RNA (dsRNA) complementary to *hTERT* mRNA sequence. GAPDH is used as a loading control. Lanes 1, 3, and 5 are non-treated HEK controls at days 1, 3, and 6 after transfection, respectively. Lanes 2, 4, and 6 are treated HEK samples at days 1, 3, and 6 after transfection, respectively. hTERT knockdown can be seen after 1 day of dsRNA treatment. The level of hTERT mRNA is depleted in the cells because of RNA interference (RNAi) and is almost recovered 6 days after treatment.

2. Materials

2.1. Cell Culture

1. HEK 293 cells (cat. no. CRL-1573, ATCC, Manassas, VA).
2. Dulbecco's modified Eagle's medium (Invitrogen Corporation, Carlsbad, CA) supplemented with 10% fetal bovine serum (10% FBS/DMEM) (HyClone, Ogden, UT).
3. Solution of trypsin (0.25%) and ethylenediamine tetraacetic acid (EDTA) (1 mM) (Invitrogen Corporation).

2.2. In Vitro Transcription of dsRNA

1. MEGAscript RNAi Kit (Ambion, Austin, TX) containing ribonucleotides (ATP, CTP, GTP, and UTP), nuclease-free water, T7 enzyme mix, and 10× T7 reaction buffer.
2. Oligonucleotide primers flanking the 200–400 bp target region to be amplified. Forward and reverse primer sequences should be designed with a T7 promoter sequence (5'-TAATACGACTCACTATAGGGAGA-3') at each 5' end.
3. Forward and reverse primers without T7 promoter sequences.
4. cDNA from HEK or an appropriate cell line that is *hTERT*-positive.

2.3. Transfection

1. Fugene 6 transfection reagent (Roche Diagnostics Corporation, Indianapolis, IN). Store at $-4\,^{\circ}$C.
2. Fully transcribed and purified dsRNA.
3. DMEM. Pre-warm at $37\,^{\circ}$C before transfection.
4. 6-Well tissue culture plates.
5. Hemacytometer.
6. Trypan blue.

3. Methods

The amount of dsRNA required for an efficient knockdown of the target gene depends on the length of the transfected transcript. The longer the transcript, the more is required to achieve adequate copies for good transfection efficiency. Therefore, it is advisable to optimize conditions such as the amount of transcript and the ratio of Fugene transfection, because the latter will differ between cell lines. It is not as necessary to test for knockdown efficacy using multiple targets because multiple siRNAs will be generated from a single dsRNA in the cell because of the effects of DICER. However, because of potential non-specific effects due to RNAi of other gene targets, the selected target sequence should be checked for sequence homology to other gene transcripts. Sequence comparisons and searches can be run on the Blast search program, which can be found at http://www.ncbi.nlm.nih.gov.

3.1. In Vitro Transcription of dsRNA

Follow according to manufacturer's instructions using the MEGAscript RNAi Kit (*see* **Note 1**).

1. Prepare the DNA template for long dsRNA synthesis by PCR amplifying the target region using appropriate cDNA. Two PCRs are to be performed. One to generate the sense-strand DNA template, and the other to generate the anti-sense-strand DNA template. To generate a reaction for the transcription of the sense strand, use the forward T7 promoter primer and reverse primer. To generate a reaction for the transcription of the anti-sense strand, use the reverse T7 promoter primer and forward primer (*see* **Fig. 2**, steps 1 and 2). The PCR conditions are based on designed primer sequences.
2. For each separate PCR, perform in vitro synthesis of ssRNA by adding $4\,\mu$l PCR DNA template to $2\,\mu$l each of ribonucleotides (ATP, CTP, GTP, and UTP), $4\,\mu$l nuclease-free water, $2\,\mu$l T7 enzyme mix, and $2\,\mu$l $10\times$T7 reaction buffer. Add the reaction buffer last to prevent precipitation.
3. Incubate both reactions at $37\,^{\circ}$C for $16\,$h (*see* **Fig. 2**, step 3).

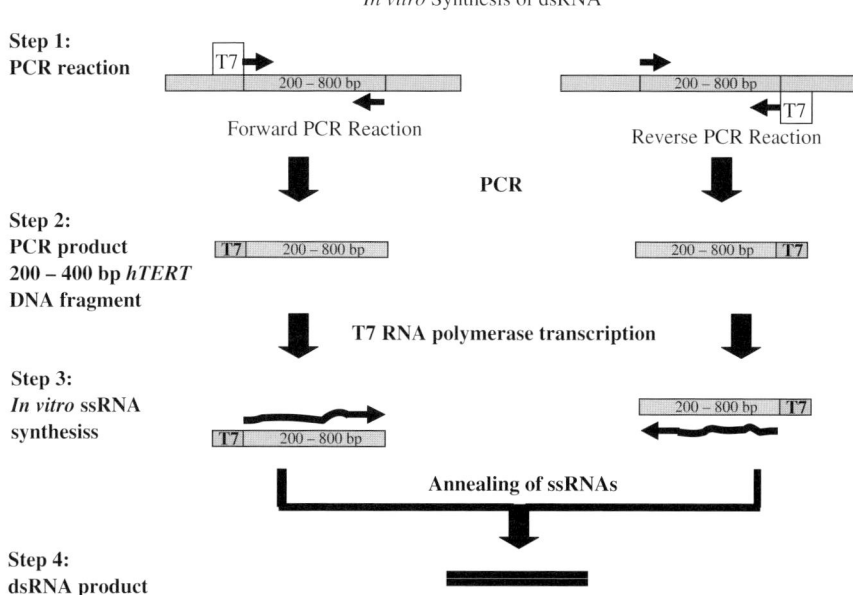

Fig. 2. Flow chart of in vitro synthesis of double-stranded RNA (dsRNA). Primers with T7 promoters are used in separate PCRs to obtain amplified products of 200–800 bp target DNA fragments flanked with T7 promoters. Sense and anti-sense single-stranded RNAs (ssRNAs) are synthesized by T7 RNA polymerase transcription, and the ssRNAs are allowed to anneal to form dsRNA.

4. To generate dsRNA, mix the two reactions together and incubate at 75 °C for 5 min. Let cool to room temperature to allow the ssRNA to slowly anneal (*see* **Fig. 2**, step 4).

3.2. HEK Transfection

1. HEK 293 cells from ATCC are cultured with 10% FBS/DMEM at 37 °C and 5% CO_2. They are maintained in 10-cm tissue culture dishes and passaged with trypsin/EDTA at 80% confluency to maintain log phase of cell growth (especially for contact-inhibited cell lines), because good transfection efficacy depends to a large degree on the cells being in log phase. Antibiotics should not be added to the culture medium during transfection with Fugene 6 transfection reagent because the influx of antibiotics with the transfection reagent is highly toxic to the cells. All cell culture and transfection procedures are performed in a sterile hood.
2. 24 hours before transfection, the cells are trypsinized and viable cells counted with a hemacytometer using the trypan blue dye exclusion method. 4.0×10^5 cells are

plated per well of a 6-well plate and incubated for 24 h to provide experimental cultures of approximately 60% confluency for transfection the next day (*see* **Note 3**).

3. 2.5–3 μg of dsRNA is required per well. Fugene 6 transfection reagent is added to pre-warmed DMEM (37°C, without serum) at a ratio of 1:5 (microgram dsRNA : microliter Fugene) carefully into the middle of the microfuge tube to prevent Fugene 6 transfection reagent from contacting and sticking to the walls of the tube. Carefully pipette the dsRNA into the middle of the tube. Incubate the mixture unshaken for at least 15 min at room temperature (*see* **Note 4**).

4. Vortex the transfection mixture briefly to obtain an even suspension and add the appropriate amount around the well in a drop-wise manner. Incubate and harvest as needed for analysis of knockdown effects.

4. Notes

1. Commercial kits for in vitro synthesis of dsRNA are also available from other vendors. A review of various methods to synthesize dsRNA can be found in **ref. 13**.

2. Generation of dsRNA longer than 800 bp may require different conditions than stated here. Refer to the manufacturer's instructions in the Ambion manual.

3. Media does not need to be replaced after transfection. However, it is advisable to provide the cells with fresh media 24 h before transfecting to ensure that cells have adequate nutrients for the incubation period following transfection. Replacing media after transfection may cause a quicker recovery from RNAi.

4. Transfection efficiency may also be compromised if RNA is not pure. To check for purity, the ratio of the optical density at wavelength 260 and 280 nm in a UV spectrophotometer should be determined. A pure sample of RNA would yield a A260/A280 ratio of 1.9–2.1.

References

1. Baglioni, C., and Nilsen, T. W. (1983) Mechanisms of antiviral action of interferon. *Interferon* **5:** 23–42.

2. Plasterk, R. H. (2002) RNA silencing: the genome's immune system. *Science* **296:** 1263–1265.

3. Yang, S., Tutton, S., Pierce, E., and Yoon, K. (2001) Specific double-stranded RNA interference in undifferentiated mouse embryonic stem cells. *Mol. Cell. Biol.* **21:** 7807–7816.

4. Elshabir, S. M., Lendeckel, W., and Tuschl, T. (2001) RNA interference is mediated by 21- and 22-nucleotide RNAs. *Genes Dev.* **15(2):** 188–200.

5. Hutvágner, G., and Zamore, P. D. (2002) RNAi: nature abhors a double-strand. *Curr. Opin. Genet. Dev.* **12(2):** 225–232.

6. Elbashir, S. M., Harborth, J., Lendeckel, W., Yalcin, A., Weber, K., and Tushcl, T. (2001) Duplexes of 21-nucleotide RNAs mediate RNA interference in cultured mammalian cells. *Nature* **411:** 494–498.

7. Snove, O., Jr., and Holen, T. (2004) Many commonly used siRNAs risk off-target activity. *Biochem. Biophys. Res. Commun.* **319(1):** 256–263.
8. Karpala, A. J., Doran, T. J., and Bean, A. G. (2005) Immune responses to dsRNA: implications for gene silencing technologies. *Immunol. Cell Biol.* **83(3):** 211–216.
9. Ambion (n.d.) Retrieved April 4, 2007. http://www.Ambion.com.
10. Shay, J. W., and Gazdar, A. F. (1997) Telomerase in the early detection of cancer. *J. Clin. Pathol.* **50:** 106–109.
11. Ahmed, A., and Tollefsbol, T. O. (2003) Telomeres, telomerase and telomerase inhibition: clinical implications for cancer. *J. Am. Geriatr. Soc.* **51:** 116–122.
12. Herbert, B. S., Wright, A. C., Passons, C. M. et al. (2001) Effects of chemopreventive and antitelomerase agents on the spontaneous immortalization of breast epithelial cells. *J. Natl. Cancer Inst.* **93:** 39–45.
13. Berletch, J. B., Green, J., Cuningham, A. P., Andrews, L. G., and Tollefsbol, T. O. (2004) Novel approaches for RNA interference and their application in cancer therapy. *Curr. Pharmacogenomics* **2:** 313–324.

4

RNA Interference Using a Plasmid Construct Expressing Short-Hairpin RNA

Serene R. Lai, Lucy G. Andrews, and Trygve O. Tollefsbol

Summary

RNA interference (RNAi) is one of the most commonly used procedures for gene targeting in today's cutting edge technology and has great potential for use in clinical therapy. Using a plasmid construct that exogenously expresses short-hairpin RNAs (shRNAs) targeting a desired gene transcript not only helps to study the downstream effects of a gene product but also offers an alternative to viral vectors for gene therapy. Using a plasmid vector to knockdown a gene allows for long-term and permanent gene knockdown, without the need to generate knockout genotypes. Here, we detail the methodology for constructing a plasmid targeting the human telomerase reverse transcriptase (*hTERT*) gene through RNAi using the Ambion pSilencer system.

Key Words: Plasmid; RNAi; shRNA; human embryonic kidney cells; hTERT.

1. Introduction

Generating knockout genotypic cell lines can be difficult and time consuming. Since the discovery of RNA interference (RNAi), knocking down a gene to prevent the translation of its transcript is an easier alternative to study the downstream effects of that gene as compared with the more traditional methods of gene targeting. RNAi can be performed in multiple ways. Here, we show the method of knocking down human telomerase reverse transcriptase (*hTERT*) gene using a plasmid construct that expresses a transcript that forms a short-hairpin RNA (shRNA).

From: *Methods in Molecular Biology, vol. 405: Telomerase Inhibition*
Edited by: L. G. Andrews and T. O. Tollefsbol © Humana Press Inc., Totowa, NJ

One way to decrease the amount of non-specific knockdown effects of RNAi *(1,2)* is to construct a short 19-mer target of the mRNA from the desired gene. The use of a shorter target sequence reduces the chance of sequence homology between the target transcript and other gene transcripts as compared with the use of long double-stranded RNA. Multiple targets should be selected because the effects of RNAi are variable depending on target sequences, cell types, and mode of delivery into the cells. Another important factor is the structure of the target sequence mRNA, which affects accessibility of the RNA-induced silencing complex (RISC) or the binding of the complementary shRNA *(3)*.

The gene-specific insert is constructed to transcribe a hairpin structure consisting of a 19 bp sequence from the target transcript, a spacer of 9 bp of random sequence for the hairpin loop, followed by the reverse complement of the original 19 bp sequence and a RNA polymerase II termination sequence consisting of a series of thymine nucleotides *(4)* (*see* **Fig. 1**). The plasmid construct should include either a H1 or a U6 promoter element at the 5′ end of the insert to allow RNA polymerase II recognition and binding, for the transcription of the insert. Intrinsically, RNA polymerase II transcribes mRNA and small nuclear RNAs. To reduce off-target effects, the 9 bp random hairpin sequence should contain no sequence homology to any gene transcript, and these sequences can be obtained from several commercial sources *(5)* or by checking its homology to other transcripts at http://www.ncbi.nlm.nih.gov by using the blast search.

The use of plasmid constructs for stable RNAi is advantageous over retroviral constructs because there is no need for concern of random integration into the genome. However, there is a risk that the cells may tend to exclude the plasmid vector after long-term cell culture. The long-term knockdown of *hTERT* is a useful tool in analyzing the role of telomerase on differentiating or

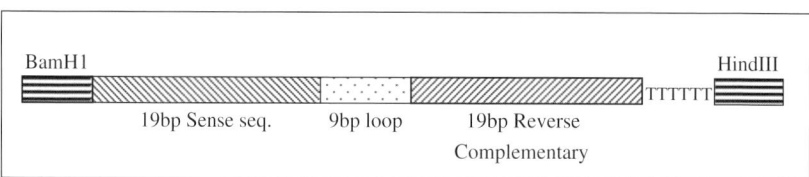

Fig. 1. Construct of a typical short-hairpin RNA (shRNA) vector insert, 5′–3′. Different restriction sequences are placed on the 5′ and 3′ ends. A 19 bp sequence for the target mRNA (sense sequence), 9 bp stem loop, and a 19 bp reverse complementary of the target sequence. When transcribed, the insert will form a secondary hairpin structure to cause the knockdown of the target gene.

cancerous systems, because telomerase activity is up-regulated in most cancer cell types. As telomeres are also thought to act as a biological clock, inducing the repression of hTERT by knocking the gene down using plasmid RNAi can prove useful in studying the developmental and aging aspects of different cell types.

2. Materials

2.1. Cell Culture

1. HEK 293 cells (cat. no. CRL-1573, ATCC, Manassas, VA).
2. HEK medium: Dulbecco's modified Eagle's medium (DMEM) (Invitrogen Corporation, Carlsbad, CA) supplemented with 10% fetal bovine serum (FBS) (HyClone, Ogden, UT).
3. Solution of trypsin (0.25%) and ethylenediamine tetraacetic acid (EDTA) (1 mM) (Invitrogen Corporation).

2.2. Plasmid Construct

1. pSilencer 3.1 H1-neo kit (Ambion, Austin, TX), containing plasmid with H1 promoter and neomycin/ampicillin antibiotic resistance.
2. Oligonucleotide sense sequence: *Bam*H1 restriction site/19-mer target sequence/9 bp loop/19-mer reverse complementary sequence/TTTTT/*Hind*III restriction site.
4. Oligonucleotide antisense sequence.
5. Restriction endonucleases *Hind*III 10 U/µl and *Bam*H1 10 U/µl (Roche Diagnostics Corporation, Indianapolis, IN).
6. Restriction digestion buffer: 10× SURE cut buffer B (Roche Diagnostics Corporation) [100 mM Tris–HCl (pH 8.0), 50 mM $MgCl_2$, 1000 mM NaCl, 10 mM 2-mercaptoethanol].
7. T4 DNA ligase, 3 U/µl (Promega Corporation, Madison, WI).
8. 10× T4 DNA ligase buffer (Promega Corporation) [300 mM Tris–HCl (pH 7.8), 100 mM $MgCl_2$, 100 mM dithiothreitol (DTT), 10 mM ATP].
9. QIAquick DNA cleanup systems (Qiagen, Valencia, CA).

2.3. Bacterial Culture

1. Max efficiency DH5α competent *Escherichia coli* cells (Invitrogen Corporation).
2. Luria Broth medium, ampicillin 100 µg/ml.
3. Luria Broth medium agar plates, ampicillin 100 µg/ml.
4. Ligated plasmid with insert.
5. QIAprep Spin Miniprep Kit (Qiagen).

2.4. Transfection and Selection

1. Fugene 6 transfection reagent (Roche Diagnostics Corporation). Store at −4 °C.
2. Purified plasmid clones.

3. DMEM. Pre-warm at 37 °C before transfection.
4. 100-mm tissue culture plates.
5. HEK selection medium: DMEM, 10% FBS, 300 µg/ml neomycin G418 antibiotic (cat. no. A1720, Sigma, St. Louis, MO).

3. Methods

This protocol is specially designed for the construction of a plasmid vector using the pSilencer system from Ambion (*see* **Note 1**). There are, however, other compatible plasmids available for RNAi on the market such as fluorescence-producing or inducible systems (*see* **Notes 2** and **3**). It should be noted that many commercially obtained plasmids may be sold as linearized plasmids, but it is always good practice to ensure that the plasmids are linearized and do not contain any foreign inserts. This can be done by digesting with the appropriate restriction enzymes and purifying the plasmid.

After selecting and picking clones, it is advisable to check for the presence of the insert by PCR amplification or DNA sequencing. DNA sequencing would also determine whether the insert contains any mutations because of cloning artifacts.

3.1. Annealing Oligonucleotides

According to the manufacturer's instructions,

1. Anneal equal volumes (2 µl each) of the sense and antisense siRNA target oligonucleotides at a concentration of 1 µg/µl each with 46 µl 1× DNA annealing solution provided with the pSilencer 3.1 H1-neo kit.
2. Incubate at 90 °C for 3 min to ensure that the oligonucleotides are not folding on themselves.
3. Incubate at 37 °C for 1 h.
4. Prepare a working stock of annealed oligonucleotides at a concentration of 8 ng/ml.
5. Store at −20 °C until ready for ligation.

3.2. Cloning into pSilencer Plasmid

1. Digest 1.5–2 µg purified plasmid with 0.2 µl BamH1 (10 U/µl), 0.2 µl HindIII (10 U/µl), and 2.5 µl 10× SURE cut buffer B and sterile deionized water to a final volume of 25 µl. Add the enzymes last after mixing the solution gently to ensure that the enzymes do not denature because of improper salt conditions (*see* **Note 4**).
2. Digest the mixture at 37 °C for 1 h and then inactivate the enzymes by incubating at 65 °C for 15 min.
3. Purify the digested plasmid with the QIAquick DNA Cleanup Systems kit. Follow the Nucleotide Removal Kit Protocol and elute with 20 µl Elution buffer (EB) [10mm Tris-ce, pH 8.5] buffer. Store at −20 °C until ready for ligation.

4. To clone the annealed insert into the plasmid, add 4 μl diluted insert, 1.33 μl nuclease-free water, 2 μl digested plasmid, 1 μl 10× T4 DNA ligase buffer, 1.67 μl T4 DNA ligase. Be sure to add the T4 DNA ligase last.
5. Ligate at 15 °C for 16 h.
6. Store at −20 °C until ready for transformation.

3.3. Transformation

1. Thaw DH5α competent *E. coli* cells on ice.
2. To 25 μl aliquots of competent cells, add 4.5 μl ligation reaction and mix by gently flicking the microfuge tube.
3. Incubate the mixture on ice for 30 min.
4. Place 37 °C water bath for 45 s for heat shocking.
5. Incubate on ice for 2 min, being careful not to mix or shake the mixture.
6. Add cells to 950 μl pre-warmed Luria Broth medium with ampicillin.
7. Incubate in a 37 °C shaker for exactly 1 h. Secure the microfuge tube horizontally in the shaker to ensure that the cells receive adequate oxygen for maximal growth.
8. Plate the transformed cells on Luria Broth plates with ampicillin. It is useful to inoculate 100 and 200 μl the reaction to obtain enough isolated clones. The remaining transformed cells may be kept at 4 °C for further inoculation on Luria Broth agar plates if necessary.
9. Incubate the plates inverted at 37 °C overnight.

3.4. Clone Amplification

1. Pick isolated colonies and inoculate each clone into 3 ml aliquots of Luria Broth with ampicillin.
2. Incubate in a 37 °C shaker for 16 h. A glycerol stock of the individual bacterial clones can be made and stored for future use (*see* **Notes 5** and **6**).
3. Pellet the cells by centrifugation. Aspirate the supernatant.
4. Wash 1× with 700 μl phosphate-buffered saline (PBS) and pellet by centrifugation. Aspirate the supernatant.
5. Store at −20 °C until ready for miniprep.
6. Follow the manufacturer's instructions using the QIAspin Miniprep Kit and elute with 50 μl sterile deionized water. It is important to use water for elution instead of the provided buffer from the kit because it might interfere with PCR or DNA sequencing procedures.
7. Each clone can be sent for DNA sequencing or PCR to confirm the presence of the insert.

3.5. HEK Cell Culture and Transfection

1. HEK 293 cells obtained from ATCC are cultured with HEK medium at 37 °C and 5% CO_2. They are maintained in 100-mm tissue culture dishes and passaged with trypsin/EDTA at 80% confluency to maintain log phase of cell growth (especially

for contact-inhibited cell lines), because transfection efficacy depends highly on the phase of cell growth. HEK cells are usually split 1:10–1:12 with each passage.

2. HEK cell cultures should be no more than 60% confluency on the day of transfection.
3. If antibiotics are used for cell-culture maintenance, the culture should be washed 1× with pre-warmed HEK medium without antibiotics and replaced with fresh medium before transfection (*see* **Note 7**). Antibiotics should not be added to the culture medium during transfection with Fugene 6 transfection reagent because the influx of antibiotics with the transfection reagent into the cells is highly toxic.
4. 10–11 µg plasmid is required per 100-mm dish. Fugene 6 transfection reagent is added to pre-warmed DMEM (37 °C, without serum) at a ratio of 1:5 (microgram : microliter Fugene). Add the transfection reagent carefully into the middle of the microfuge tube to avoid the reagent from contacting and sticking to the walls of the tube. The appropriate amount of plasmid is carefully pipeted into the middle of the tube and should appear as a clear liquid in a contained suspension with the Fugene 6 transfection reagent. Incubate the mixture unshaken for at least 15 min at room temperature.
5. Vortex the transfection mixture briefly to obtain an even suspension and add the appropriate amount around the well in a drop-wise manner. Incubate at 37 °C for 24–48 h.
6. Replace the HEK medium with HEK selection medium. Maintain and refresh the HEK selection medium every 3 days. HEK G418-resistant cells will persist and proliferate (*see* **Note 8**). Split the cell culture as needed.

4. Notes

1. Plasmids with polymerase III promoters H1 or U6 can be used in the production of RNA transcripts because they have well-defined transcription start and termination sequences and do not produce poly-adenosine tails in their transcripts *(6)*.
2. It is sometimes useful to have a plasmid system that also confers green fluorescent protein (GFP) fluorescence to confirm the presence of the plasmid in the cell, especially for studies that involve observing phenotypic changes of targeted cells.
3. Temperature sensitive or other inducible plasmid systems (such as tetracycline) may be used if the knockdown of a gene is toxic or fatal to the cells.
4. When adjusting the conditions for multiple restriction enzyme digestion, ensure that the digestion buffer is compatible for the usage of all enzymes. Also, the total volume of the enzymes should not exceed 10% of the total volume of the reaction because of the fact that the glycerol content in the restriction enzyme stock may interfere with the digestion process.
5. Store transformed bacteria with desired clones in 15% glycerol/cell suspension in Luria Broth. Snap freeze in liquid nitrogen and store at −80 °C.
6. When growing bacteria in Luria Broth suspension, it is not advisable to inoculate directly from a frozen stock because of purity and yield concerns. Instead, streak on Luria Broth agar and incubate for 24 h at 37 °C, then inoculate from an isolated clone.

7. Washing HEK cells on the culture plate should only be done with pre-warmed HEK medium. Using PBS will only cause the cells to lose adherence and lift off the plate.

8. Selection of transfected HEK cells on HEK selection media should continue for at least 2 weeks before the cells may be harvested to ensure thorough selection of cells.

References

1. Elbashir, S. M., Harborth, J., Lendeckel, W., Yalcin, A., Weber, K., and Tushcl, T. (2001) Duplexes of 21-nucleotide RNAs mediate RNA interference in cultured mammalian cells. *Nature* **411:** 494–498.

2. Snove, O., Jr., and Holen, T. (2004) Many commonly used siRNAs risk off-target activity. *Biochem. Biophys. Res. Commun.* **319:** 256–263.

3. Overhoff, M., Alken, M., Kretschmer-Kazemi Far, R., Lemaitre, M., Lebleu, B., Sczakiel, G., and Robbins, I. (2005) Local RNA target structure influences siRNA efficacy: a systematic global analysis. *J. Mol. Biol.* **348(4):** 871–881.

4. Brummelkamp, R. T., Bernards, R., and Agami, R. (2002) A system of stable expression of short interfering RNAs in mammalian cells. *Science* **296:** 550–553.

5. Ambion (n.d.) SiRNA design guidelines. Retrieved April 4, 2007. http://www.ambion.com/techlib/tb/tb_506.html.

6. Ill, C. R., and Chiou, H. C. (2005) Gene therapy progress and prospects: recent progress in transgene and RNAi expression cassettes. *Gene Therapy* **12:** 795–802.

5

Retrovirus-Mediated RNA Interference
Targeting hTERT Through Stable Expression of Short-Hairpin RNA

Amanda P. Cunningham, Lucy G. Andrews, and Trygve O. Tollefsbol

Summary

RNA interference (RNAi) has recently emerged as a reliable tool for studying the effects of knocking down or ablating the expression of specific genes. It is hoped that progress made in the laboratory toward in vitro down regulation of gene expression may be carried over into the clinic for treatment of diseases in which the expression of a specific gene is associated with initiation or progression of that disease. Such is the case with telomerase, an exciting drug target that has been the focus of numerous investigations with a wide variety of inhibitors. This chapter describes the use of retrovirally introduced short-hairpin RNA as an effector of stable, long-term RNAi in human cells.

Key Words: RNA interference; retroviral; short-hairpin RNA; telomerase inhibition.

1. Introduction

RNA interference (RNAi) is rapidly becoming the technique of choice for gene knockdown and analysis of function studies. There is hope that the technology could eventually be used as a treatment for diseases in which the expression of a specific gene is associated with disease progression. The catalytic subunit of telomerase, human telomerase reverse transcriptase (hTERT), seems to be a logical target for RNAi because advances in telomerase inhibition have considerable potential for cancer therapy.

There are numerous approaches to stable RNAi, both plasmid-based and virally delivered. Several laboratories have designed their own vectors (*1–7*) for

From: *Methods in Molecular Biology, vol. 405: Telomerase Inhibition*
Edited by: L. G. Andrews and T. O. Tollefsbol © Humana Press Inc., Totowa, NJ

short-hairpin RNA (shRNA) delivery *(8)* and have found them to be effective. Many companies market some RNAi product, and our laboratory has chosen the GeneSuppressor™ System (cat. no. IMG-1000) from Imgenex for retroviral delivery of shRNA to target cells. Retroviral delivery has the advantage of stable, long-term production of shRNA and the capacity for antibiotic selection of infected cells. One should choose an shRNA delivery system that best meets the needs of the individual investigator (ease of use, hands-on time, product availability, cells to be infected, etc.).

Although some companies offer in-house target selection for RNAi, it is often desirable to choose one's own target sequence on the hTERT mRNA. Oligonucleotide design will depend largely on the particular system being used. The size of the homologous region being targeted, and the length of the hairpin loop will vary according to vector manufacturer instructions and with individual preferences. For the purposes of this chapter, designs reflect GeneSuppressor™ System guidelines.

For the most part, RNAi target selection is still trial and error with only a few general criteria to guide the process. Several targets should be chosen and tested to find the one that yields the most complete knockdown of the hTERT mRNA. Short-interfering RNAs (siRNAs) are 21-nucleotide to 23-nucleotide duplexes of double-stranded RNA that actually mediate the RNAi process *(9)*. It is advisable to initially design siRNAs against the target sequences and test them to rapidly assess target choices before investing valuable time in vector construction and antibiotic selection. Regions of the hTERT mRNA possessing a strong secondary structure should be avoided as targets as they can be relatively inaccessible to components of the RNAi machinery. Elbashir et al. *(10)* have provided some general guidelines that are helpful in planning oligonucleotide design: 5′-untranslated and 3′-untranslated regions should be avoided as well as the region immediately downstream of the start codon. The ideal sequence for synthetic siRNAs is 5′-AA(N19)UU-3′ where N is any nucleotide. Oligonucleotides should have roughly 50% GC content *(10)*. (The AA and UU dimers in the sequence are not necessary when using the GeneSuppressor™ System's expression vector produced by Imgenex.) The motif corresponding to the coding region of hTERT should be 19-nucleotide to 21-nucleotide long, followed by a spacer of four to nine nucleotides that will form the hairpin loop. The size of the loop is important; it has been demonstrated that a loop of nine nucleotides incorporated into the shRNA design may be much more effective than a somewhat shorter (e.g., seven nucleotides) loop and therefore has a significant impact on the efficiency of the shRNA in knocking down the target mRNA *(3)*. Following the hairpin loop is the inverted complement of the motif. Oligonucleotides will need to be ordered in both the sense and the antisense configurations and annealed in the laboratory. Finally, avoid incorporating continuous stretches of adenines or thymines in the target sequence, as these may act as internal terminators. *See* **Note 4** for an example of oligonucleotide design.

2. Materials
2.1. Retroviral Construct Preparation

1. Double-stranded oligonucleotides with homology to desired target region of hTERT mRNA.
2. Retroviral vector, pSuppressorRetro™ , and packaging plasmid (Imgenex, San Diego, CA). Include retroviral LacZ plasmid if desired (*see* **Note 1**).
3. Primer annealing buffer (included in GeneSuppressor™ System, Imgenex).
4. T4 DNA ligase and 10× ligase buffer.
5. DH5α competent *Escherichia coli* (Invitrogen, Carlsbad, CA).
6. LB Levia-Bertani plates containing 100 μg/ml ampicillin.
7. LB broth containing 100 μg/ml ampicillin.
8. Miniprep and maxiprep DNA purification kits (Qiagen, Valencia, CA).

2.2. Cell Culture, Transfection, and Infection

1. HEK 293 cells (cat. no. CRL-1573, American Type Culture Collection, Manassas, VA).
2. Dulbecco's modified Eagle's medium (DMEM) (Mediatech, Herndon, VA) supplemented with 10% heat-inactivated fetal bovine serum (HyClone, Logan, UT).
3. 0.25% trypsin and 2.21 mM ethylenediamine tetraacetic acid (EDTA) (Mediatech).
4. Target cells of interest, culture medium, and dissociation reagents.
5. FuGENE 6 transfection reagent (Roche, Indianapolis, IN) (*see* **Note 2**).
6. SteriFlip 50-ml filters and 0.45-μm pore size (Millipore, Billerica, MA) (*see* **Note 3**).
7. Polybrene (hexadimethrine bromide; Aldrich, Milwaukee, WI), stock solution 8 mg/ml in sterile, molecular biology-grade water.
8. Geneticin (G418) (Invitrogen).
9. In Situ β-Galactosidase Staining Kit for monitoring transfection efficiency (Stratagene, La Jolla, CA).

3. Methods
3.1. Preparation of Oligonucleotides

1. Choose the target region of hTERT mRNA and design the DNA oligonucleotides, keeping in mind the aforementioned guidelines.

2. The pSuppressor vector from Imgenex is supplied linearized with *Sal*I and *Xba*I ends. Therefore, compatible restriction sites should be incorporated into the oligonucleotide design: *Xho*I at the 5′ end and *Xba*I at the 3′ end of the oligonucleotide sequences.
3. The 3′ end of the sense oligo should end in a sequence of five thymidines that serve as a terminator.
4. *See* **Note 4** for an example of oligonucleotides for shRNA techniques.

3.2. Retroviral Vector Construction

3.2.1. Anneal Oligos

1. Anneal the oligonucleotides in a reaction containing 1 μg each of the sense and antisense oligos, 2 μl 10× annealing buffer (*see* **Note 5**) and deionized water to a final volume of 20 μl.
2. Heat the reactions to 95 °C for 10 min, then cool the reactions *gradually* to room temperature.

3.2.2. Ligate Oligos into Vector

1. Ligate annealed oligo into linearized Imgenex pSuppressor vector in the presence of 2 μl 10× ligase buffer, 1 μl T4 DNA ligase, and deionized water to a final volume of 20 μl (*see* **Note 6**).
2. Ligation conditions: 16 °C for 12–16 h.
3. Transform DH5α competent *E. coli* using 3 μl ligation reaction. Grow transformation reaction overnight on LB plates containing 100 μg/ml ampicillin.
4. Pick several colonies to verify the presence of recombinants. Grow in LB broth in the presence of 100 μg/ml ampicillin and purify using Qiagen's Spin Miniprep Kit.
5. Perform restriction analysis of minipreps to check for the presence of the desired insert (*see* **Note 7**).
6. Confirm construct integrity by DNA sequencing (*see* **Note 8**).
7. Scale up plasmids by performing DNA maxipreps of the desired retroviral construct, the packaging plasmid we wish to use, and the retroviral LacZ plasmid (if we wish to assess transfection and infection efficiencies).

3.3. Transfection of HEK 293 Cells

1. One day prior to transfection, plate approximately 1.5 million HEK 293 cells in a 10-cm dish for each target being tested. It is very important that the media used for the HEK 293 cells *not* contain any antibiotic/antimycotic agents. The cells should be 30–50% confluent on the day of transfection (*see* **Note 9**).
2. Transfect the cells using FuGENE following the manufacturer's guidelines. Optimization experiments should be done to determine the most effective amounts of DNA to be used as well as the ratio of FuGENE to DNA. We have found ratios of 3:2 and 5:1 to be effective in transfecting HEK 293 cells. 2 μg DNA per well of

a 6-well plate works well. Keep in mind the total DNA amount includes both the retroviral vector and the packaging plasmid.
3. Incubate the FuGENE–DNA complex mixture for 45 min at room temperature before adding dropwise to culture dishes.
4. The morning after transfection, change medium in the culture dishes to the medium of our target cells. Apparently the transfected cells produce a type of cytostatic factor; changing the medium helps to dilute this factor (*see* **Note 10**).

3.4. Infection of Target Cells

1. Plate target cells 1 day prior to infection (may do 1 day after transfection) so that they will reach a density of no more than 30–50% on the day of infection (*see* **Note 11**).
2. On the day of infection, remove the virus-containing supernatant from the HEK 293 cells and filter sterilize the supernatant through a *surfactant-free* 0.45-μm filter.
3. Infect the target cells with the virus-containing supernatant using a 1:1 ratio of supernatant to target cell medium in a final concentration of 8 μg/ml polybrene (*see* **Note 12**).
4. Allow continued virus contact with the target cells for at least 48 h before beginning G418 selection. It may be advantageous to subject the target cells to repeated rounds of infection every 6–12 h or at least once per day (*see* **Note 13**).
5. Upon completion of the desired infection period, select the target cells in G418. Include a dish of untreated cells that will be used as a selection control. When all these cells are dead, selection is complete and the G418 can be removed from the medium. Optimum G418 concentration is cell type dependent and varies widely. A good rule is that there should be no noticeable cell death 2 days after adding the G418 and approximately 30–50% cell death 4 days after adding the G418. Selection should be complete in 7–10 days.
6. Maintaining the cultures will require concentration of cells into smaller dishes as they die off because of selection or addition of uninfected cells to keep cultures at a density that is conducive to maximal growth (*see* **Note 14**). Always allow cells to adhere overnight before reintroducing the G418 antibiotic.

4. Notes

1. It is strongly advised to use the LacZ plasmid to determine optimum transfection and infection conditions for our experiments. Cotransfection of the LacZ plasmid with the appropriate packaging plasmid will produce infectious viral particles that can infect the target cells and consequently be useful in assessing infection conditions. Maximizing efficiency will increase the number of infected target cells and reduce the time needed to culture sufficient cells to conduct experiments.
2. Any efficient transfection method should be adequate. The GeneSuppressor™ system from Imgenex includes its own transfection reagent and accompanying

protocol. We have found transfections with FuGENE to be highly efficient (~70% transfection efficiency).

3. Do *not* use 0.2-μm filters as they may shear the viral envelop. Use only *surfactant-free* (for the same reason) 0.45-μm filters.

4. An example of the sense and antisense oligonucleotides against the target pursued by Masutomi et al. *(11)*, including *Xho*I and *Xba*I overhang sites:

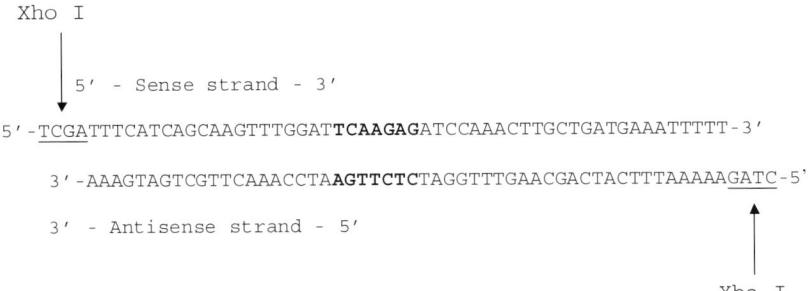

```
Xho I

       5' - Sense strand - 3'

5'-TCGATTTCATCAGCAAGTTTGGATTCAAGAGATCCAAACTTGCTGATGAAATTTTT-3'

   3'-AAAGTAGTCGTTCAAACCTAAGTTCTCTAGGTTTGAACGACTACTTTAAAAAGATC-5'

   3' - Antisense strand - 5'

                                                    Xba I
```

Restriction site overhangs are underlined. Bases comprising the loop structure are in bold. Note the 5T termination signal at the 3'-end of the sense strand. This sequence is based on a published target for RNAi of hTERT with modifications enabling use in the pSuppressor vector *(11)*.

```
Demonstration of how oligo        5'-TCGATTTCATCAGCAAGTTTGGAT T  C
folds back on itself to form                                   A
hairpin (sense strand).                                         A
Hairpin in bold type.             3'-TTTTTAAAGTAGTCGTTCAAACCTAG A  G
```

5. 10× annealing buffer: 100 mM Tris–HCl (pH 7.5), 1 M NaCl, and 10 mM EDTA (http://www.protocol-online.org/prot/Detailed/3369.html).

6. The manufacturer recommends 1 μl each of the vector and annealed oligo; we have obtained better results using a 3:1 ratio of annealed oligo to pSuppressor vector.

7. As the *Sal*I site is destroyed upon successful cloning of the insert, digestion with this enzyme should have no effect. It is helpful to digest with another enzyme in a double digest (such as *Eco*RI). Double digestion with *Sal*I and *Eco*RI will cut out an approximately 1.5 kb fragment from negative clones while linearizing positive clones at the *Eco*RI site.

8. The sequencing of some clones may not be possible because of the strong secondary structure formed by the short hairpin *(1)*. The facility rendering sequencing support

should be notified of the nature of the construct so that conditions might be optimized for sequencing through secondary structure.

9. HEK 293 cells may be maintained in culture with antibiotic/antimycotic selection but the media used during the transfection should not contain any antibiotic/antimycotic agents, as its presence can drastically reduce transfection efficiency.

10. This step is advised by Imgenex in its product manual, but the effect has not been empirically tested by our laboratory.

11. Retroviral infection is dependent upon cell division. Cells must be growing well and actively dividing to be infected.

12. It is important for the virus to be as concentrated as possible without adversely affecting the growth of the target cells. Diluting the viral supernatant 1:1 is the minimum dilution recommended by Imgenex. If, in fact, the transfected cells are producing some cytostatic factor, the dilution of the supernatant helps to reduce growth inhibition of the target cells.

13. Virus contact with target cells should be for a minimum of 12–24 h. As the half-life of the virus is only 6–8 h at 37 °C, it is advantageous to reinfect every 12 h with fresh virus. Infection for 48 h before selection is desirable.

14. Any uninfected cells that are added must also be eliminated by G418 selection. The most effective method would be to concentrate the cells into smaller dishes. However, using this technique on some cell lines may not be feasible (such as those with low reattachment efficiency).

References

1. Devroe, E. and Silver, P.A. (2002) Retrovirus-delivered siRNA. *BMC Biotechnol.* **2**, 15.
2. Zhou, H., Xia, X.G., and Xu, Z. (2005) An RNA polymerase II construct synthesizes short-hairpin RNA with a quantitative indicator and mediates highly efficient RNAi. *Nucleic. Acids Res.* **33**, e62.
3. Brummelkamp, T.R., Bernards, R., and Agami, R. (2002) A system for stable expression of short interfering RNAs in mammalian cells. *Science* **296**, 550–553.
4. Brummelkamp, T.R., Bernards, R., and Agami, R. (2002) Stable suppression of tumorigenicity by virus-mediated RNA interference. *Cancer Cell.* **2**, 243–247.
5. Sui, G., Soohoo, C., Affar el, B., Gay, F., Shi, Y., and Forrester, W.C. (2002) A DNA vector-based RNAi technology to suppress gene expression in mammalian cells. *Proc. Natl. Acad. Sci. U. S. A.* **99**, 5515–5520.
6. Stewart, S.A., Dykxhoorn, D.M., Palliser, D., Mizuno, H., Yu, E.Y., and An, D.S. (2003) Lentivirus-delivered stable gene silencing by RNAi in primary cells. *RNA* **9**, 493–501.
7. Liu, C.M., Liu, D.P., Dong, W.J., and Liang, C.C. (2004) Retrovirus vector-mediated stable gene silencing in human cell. *Biochem. Biophys. Res. Commun.* **313**, 716–720.

8. Paddison, P.J., Caudy, A.A., Bernstein, E., Hannon, G.J., and Conklin, D.S. (2002) Short hairpin RNAs (shRNAs) induce sequence-specific silencing in mammalian cells. *Genes Dev.* **16**, 948–958.

9. Elbashir, S.M., Harborth, J., Lendeckel, W., Yalcin, A., Weber, K., and Tuschl, T. (2001) Duplexes of 21-nucleotide RNAs mediate RNA interference in cultured mammalian cells. *Nature* **411**, 494–498.

10. Elbashir, S.M., Harborth, J., Weber, K., and Tuschl, T. (2002) Analysis of gene function in somatic mammalian cells using small interfering RNAs. *Methods* **26**, 199–213.

11. Masutomi, K., Yu, E.Y., Khurts, S., Ben-Porath, I., Currier, J.L., and Metz, G.B. (2003) Telomerase maintains telomere structure in normal human cells. *Cell* **114**, 241–253.

6

Telomerase Inhibition and Telomere Targeting in Hematopoietic Cancer Cell Lines with Small Non-Nucleosidic Synthetic Compounds (BIBR1532)

Hesham El Daly and Uwe M. Martens

Summary

Telomere maintenance has been shown to be essential for unlimited growth potential of human cells and is regarded as one hallmark of cancer. Telomere repeats at the ends of eukaryotic chromosomes are synthesized by the enzyme telomerase, which is active in most cancers and to some extend also in normal somatic cells. Therefore, targeting the telomerase/telomere complex offers great potential for the development of novel anticancer therapeutics. An example of such a strategy is the small molecule BIBR1532 that is a selective, non-nucleosidic inhibitor of the catalytic component hTERT. Treatment of cancer cells with this compound leads to progressive telomere shortening, consecutive telomere dysfunction, and finally growth arrest after a lag period that is largely dependent on initial telomere length. We have additionally shown that using this class of telomerase inhibitor at higher concentrations exerts a direct cytotoxic effect on malignant cells of the hematopoietic system but not on normal stem cells, which appears to derive from direct damage to the structure of individual telomeres.

Key Words: Telomerase; telomere; TRF2; telomere dysfunction; telomerase inhibition.

1. Introduction

Telomeres are specialized DNA protein structures that cap the ends of linear chromosomes and thereby protect them from fusion, exonucleolytic degradation, and aberrant chromosomal recombination *(1)*. In mammalian cells, telomeres contain repetitive double-stranded repeats of the sequence TTAGGG

From: *Methods in Molecular Biology, vol. 405: Telomerase Inhibition*
Edited by: L. G. Andrews and T. O. Tollefsbol © Humana Press Inc., Totowa, NJ

and terminate with a single-stranded 3′ extension of the G-rich strand. The capping function is mediated by specialized architecture in which the 3′ overhang participates with telomere-binding proteins in a large loop structure called T-loop (2). Telomerase is the enzyme that is responsible for maintenance of the telomere structure using the 3′ overhang as primer for de novo synthesis of telomere repeats (3). This enzyme is a ribonucleoprotein reverse transcriptase composed of an RNA template (hTR) and a catalytic protein subunit (hTERT). In most normal somatic cells, telomerase activity is absent, and telomere repeats are lost with cell division and with aging. Limited telomerase activity has also been found in stem cells and lymphocytes (4). In contrast, about 80–90% of cancer cells have detectable telomerase activity, which leads to the stabilization of telomeres and confers immortality or unlimited growth potential (5).

As most cancer cells are reliant on telomerase for their survival, telomerase has become an attractive target for the development of new cancer therapeutics (6). Targeting the active site of hTERT has been reported using some small nucleoside analogues such as AZT (3′-azido-2′, 3′-dideoxythymine), but lack of selectivity for telomerase has limited this approach (7). A novel structural class of non-peptidic, non-nucleosidic inhibitors of human telomerase has been described by the company Boehringer Ingelheim Pharma KG (Biberach, Germany) (8) that is specific in inhibition of the catalytic activity of telomerase (9). One example of this class of compounds, designated BIBR1532 {2-[(E)-3-naphtalen-2-yl-but-2-enoylamino]-benzoic acid}, inhibits the in vitro processivity of telomerase in a dose-dependent manner, with half-maximal inhibitory concentrations (IC_{50}) of 93 nM (9).

Long-term treatment of several cancer cell lines with BIBR1532 with concentrations of 10 μM had no effect on short-term cell viability or proliferation but induced progressive telomere shortening and a proliferation arrest after a characteristic lag period of several weeks, with hallmarks of senescence, including morphological, mitotic, and chromosomal aberrations.

We have furthermore shown that using BIBR1532 at concentrations of 30–80 μM exerts a dose-dependent direct cytotoxic effect within a few days, which was demonstrated in several leukemia cell lines and primary cells from leukemia patients but not in normal stem cells.

A time-dependent individual telomere erosion was observed, which was associated with loss of telomeric repeat binding factor 2 (TRF2) and increased phosphorylation of p53 suggesting that high-dose BIBR1532 interferes with the capping function of telomeres (10).

The changes in viability and proliferation were analyzed, using DiOC6/PI and WST-1 proliferation assay, respectively. The effects on telomere length dynamics were monitored by quantitative fluorescence in situ hybridization (FISH), using either digital fluorescence microscopy (Q-FISH) on metaphase chromosome spreads or flow cytometry (flow-FISH) on cells in suspension.

2. Materials

2.1. Cell Lines

The Jvm13 cell line represents an immortalized B-cell line from a patient with prolymphocytic leukemia, HL-60 is an AML cell line, and Nalm1 was derived from a patient with CML in blastic phase (Deutsche Sammlung von Mikroorganismen und Zellkulturen GmbH, Braunschweig, Germany).

2.2. Cell Lysis, SDS-Polyacrylamide Gel Electrophoresis, and Western Blotting

1. RIBA lysis buffer: 150 mM NaCl, 1% NP-40, 0.5% DOC, 1% SDS, 50 mM Tris–HCl, pH 8.0. Keep at 4°C.
2. Polyacrylamide gel:

    a. Resolving gel (12% gel): 4.5 ml 30% acrylamide, 4.3 ml H_2O, 2.25 ml Tris buffer, 8.8 pH, 112.5 μl 10% SDS, 25 μl TEMED, 37.5 μl 10% ammonium persulfate (APS).

    b. Stacking gel: 800 μl 30% acrylamide, 3.6 ml H_2O, 500 μl Tris buffer, 6.8 pH, 50 μl 10% SDS, 10 μl TEMED, 17 μl 10% APS.

3. Protease inhibitor stock solution: one tablet in 2 ml redistilled H_2O, store the aliquots at −20°C.
4. Resolving gel buffer: 1.87 M Tris–HCl, pH 8.8. Store at room temperature.
5. Stacking gel buffer: 1.25 M Tris–HCl, pH 6.8. Store at room temperature.
6. 10% SDS, 10 g SDS plus H_2O to 100 ml. Store at room temperature.
7. 10% APS, 1 g APS plus H_2O to 10 ml, store the aliquots at −20°C.
8. 5× running buffer, 10 g Tris, 48 g glycine, 5 g SDS plus H_2O to 1 l. Store at room temperature.
9. Transfer buffer: 3.03 g Tris, 14.41 g glycine, 200 ml methanol plus H_2O to 1 l. Store at room temperature.
10. Blocking buffer: 3 g bovine serum albumin (BSA), 3 g non-fat dry milk, make up in phosphate-buffered saline (PBS) to 100 ml. Use also for primary and secondary antibody dilution. Keep at 4°C to prevent bacterial contamination.
11. Washing buffer: 0.5 ml Tween 20 in 500 ml PBS. Store at room temperature.
12. Primary antibodies: ß-actin (Biocarta, Hamburg, Germany), hTERT (Novocastra, Newcastle, England), p53 (Biocarta), phospho p53 (Biocarta), and TRF2 (Biocarta).

13. Anti-mouse, horseradish peroxidase (Amersham Biosciences, Buckinghamshire, UK). Anti-rabbit, horseradish peroxidase (Amersham Biosciences).

14. Enhanced chemiluminescent (ECL) kit from Amersham Biosciences. Prepare the Working Reagent prior to use.

2.3. Flow-FISH for Mean Telomere Length Measurement

1. Telomere-specific peptide nucleic acid (PNA) Probe ((AATCCC)$_3$-FITC, Applied Biosystems, Foster City, CA). A stock is prepared, diluted with double distilled H$_2$O to make a concentration of 30 μg/ml, which is stored in the dark at −20°C until use (*see* **Hybridization Mix Preparation Table**).

2. Hybridization mix: 0.3 μg/ml PNA-Probe in 70% deionized formamide, 20 mM Tris–HCl, pH 7, 1% BSA. Store at −20°C till use (*see* **Note 1**). The deionized formamide (Merk, Darmstadt, Germany) and the BSA (Sigma-Aldrich, Steinheim, Germany) are aliquoted in 15-ml and 50-ml Falcon tubes, respectively, and stored at −20°C until use.

Hybridization mix		Negative control	End concentration
1 M Tris, pH 7	4	4	20 mM
10% BSA	20	20	1%
Formamide (deionized)	140	140	70%
PNA-Probe (30 μg/ml)	2	–	0.3 μg/ml
ddH$_2$O	34	36	
	200 μl	200 μl	

3. Wash I: 70% formamide, 10 mM Tris–HCl, pH 7, 0.1% BSA, 0.1% Tween 20. Prepare fresh prior to use.

Wash I	Negative control	End concentration
1 M Tris, pH 7	10	10 mM
10% BSA	10	0.1%
10% Tween 20 (in H$_2$O)	10	0.1%
Formamide	700	70%
ddH$_2$O	270	
	1000 μl	

4. Wash II: PBS + 0.1% BSA, 0.1% Tween 20. Prepare fresh prior to use.

Wash II	Negative control	End concentration
PBS + 0.1% BSA	990	
10% Tween 20 (in PBS + 0.1% BSA)	10	0.1%
	1000 μl	

5. Propidiumiodide mix: 0.05 μg/ml propidiumiodide, 10 μg/ml RNase in PBS + 0.1% BSA. Prepare fresh prior to use. The RNase DNase-free 0.5 μg/μml (Boehringer, Mannheim, Germany) stored at $-20°C$ until use (*see* **Note 2**).

Propidiumiodide mix	Negative control	End concentration
PBS + 0.1% BSA	292	
RNase (0.5 μg/μl)	6	10 μg/ml
Propidiumiodide (10 μg/ml)	1.5	0.05 μg/ml
	300 μl	

2.4. Genetic Analyses

1. Inhibitor of mitotic spindle: 0.1 μg/ml colcemide.
2. Hypotonic solution: 0.075 M KCl.
3. Fixative: methanol : acetic acid (3:1).

2.5. Q-FISH for Telomere Length Measurement

1. Hybridization mix: 70% deionized formamide (Merck, Darmstadt, Germany) stored at $-20°C$ until use, 0.25% NEN blocking solution (Perkin Elmer, Boston, MA), 10 mM Tris–HCl, pH 7.0, 0.5 μg/ml PNA tel Cy3 probe (Applied Biosystems).
2. Washing buffer I: 70% formamide, 0.1% BSA, 0.01 M Tris–HCl, pH 7–7.5.
3. Washing buffer II: 0.1 M Tris–HCl, 0.15 M NaCl, 0.08% Tween 20, pH 7–7.5.
4. Counterstain: 1 mg/ml 4,6-diamidino-2-phenylindole (DAPI), 1:100 in PBS, and from that solution, a new 1:25 dilution in Vectashield mounting media (Vector Laboratories, Peterborough, UK).
5. Digital camera for imaging (Sensys, Photometric) on an Axioplan II fluorescence microscope using the Vysis workstation QUIPS (Vysis, Downers Grove, IL). The TFL-TELO software *(11)* is used for the analysis of telomere profiles.

3. Methods

3.1. Preparation of Mononuclear Cells (Ficoll)

Pipette 9 ml peripheral ethylenediaminetetraacetic acid (EDTA) blood or 5 ml bone marrow samples into 50-ml Falcon tubes and then fill up to 50 ml with PBS. Pipette 15 ml Ficoll Separating Solution (Biochrom KG, Berlin, Germany) into another 50-ml Falcon tube and then fill slowly with the diluted samples. After 20-min centrifugation at $900 \times g$ and 18°C without brakes, throw away the supernatant. Aspirate the phase with the mononuclear cells and pipette into a new 50-ml Falcon tube and wash two times with 50 ml PBS (10 min at $300 \times g$ and 18°C). Resuspended the cells in PBS + 0.1% BSA and then count the cells.

3.2. Cell Culture

1. The cell lines are grown/plated in a culture medium, which is composed of RPMI 1640 (Invitrogen, Karlsruhe, Germany) supplemented by 10% heat-inactivated fetal calf serum (FCS) (Invitrogen), 2 mM L-glutamine, and 1% penicillin/streptomycin (complete medium). For the detailed cell lines' final plating concentration and splitting intervals, please refer to provider's recommendations. During the early passages, backup aliquots are prepared and stored at −25°C till needed.
2. Dissolve BIBR1532 (Boehringer Ingelheim Pharma KG) in 0.1% dimethyl sulfoxide (DMSO), divide in 1 mM aliquots, and store at −25°C till use. Prepare the following concentrations in the same medium as used for cell culture as diluent: 10, 30, 50, and 80 µM. Respectively, use similar concentrations of DMSO as negative controls. Add the different concentrations of BIBR1532 and DMSO to the culture medium prior to splitting the cells.
3. Incubate the cells in culture in an incubator (BBD6220, Heraeus, Stuttgart, Germany) at 37°C and 5% CO_2 under sterile conditions.
4. Cell count: count the cells using Neubauer-counting chamber, using trypan blue (Invitrogen) exclude the dead cells and calculate the cell concentrations as follows:

$$\text{Cell count/ml} = \frac{\text{Cell count}_{4\,\text{Big squares}}}{4} \times \text{Dilution factor} \times 10,000.$$

Cell growth is calculated and expressed as *cumulative population doubling level* (CPDL):

$$PD = \frac{\log \text{Cell count}_{\text{Output}} - \log \text{Cell count}_{\text{Input}}}{\log 2}, \quad CPDL = \sum PD.$$

3.3. Western Blot Analysis for TRF2 and p53

1. Collect the cells and pellet the cells using the centrifuge (Eppendorf microfuge) at $1500 \times$ g for 5 min at 4°C. Lyse the pellet with 100 μl lysis buffer on ice for 60 min, with vortexing every 10 min. Centrifuge the tubes at $16,000 \times g$ in an Eppendorf microfuge for 10 min at 4°C. Transfer the supernatant to a new tube and discard the pellet. Determine the protein concentration using Bradford assay (Bio-Rad Laboratories GmbH, Munich, Germany). Use 10–50 μg protein, adjust with distilled H_2O to 22.5 μl, then add 7.5 loading buffer to make a final volume of 30 μl. Boil the samples for 5 min, then cool at room temperature for 5 min. Flash spin the tubes to bring down condensation prior to loading gel.

2. Assemble the glass plates and spacers. Pour the resolving gel and add 70% ethanol as a seal till the gel solidifies. When the gel is set, pour the ethanol off and rinse with deionized H_2O. Pour the stacking gel and insert the comb immediately. When the stacking gel is set, immerse the chamber in $1\times$ running buffer.

3. Flash spin the samples and then load into the wells. Use 15 μl Bio-Rad Kaleidoscope Prestained Standards (cat. no. 161-0324) directly as a marker. Constant current is used (voltage set at 70 V for 40 min, then 120 V for 2 h).

4. Cut a piece of transfer membrane (mobilon-P, Millipore Corporation, Bedford, UK) and leave 30 s in methanol at room temperature, 2 min in distilled H_2O, and 15 min in transfer buffer.

5. Pre-wet the sponges and filter papers (slightly bigger than gel) in $1\times$ transfer buffer. Allow transfer for 1 h at 350 mA at 4°C. After finishing, stain membranes with ponso (red stain) and leave on the orbital shaker or rocker for about 15 min (*see* **Note 3**). Wash with H_2O and immerse in blocking buffer and leave overnight.

6. Incubate membranes with primary antibody diluted in freshly prepared blocking buffer for 2 h at room temperature or overnight at 4°C. Wash as follows: $3\times$ 10 min with 0.05% Tween 20 in PBS (*see* **Notes 4 and 5**). Dilute the secondary antibody in blocking buffer and incubate for 45–60 min at room temperature. Wash the membranes $3\times$ 10 min with 0.05% Tween 20 in PBS. Use Amersham ECL kit for detection.

7. Prepare ECL kit Working Reagent prior to use. 5 ml of Working Reagent is needed for a single membrane. Put the film into the cassette in a dark room with a developer. Incubate the film for about 1–5 min (variable). Pass the film through the developer.

8. Rinse off the blot with 0.05% Tween 20 in PBS. Add about 5–10 ml stripping buffer and incubate for 15 min. Pour off the solution and wash the membranes for $3 \times 0.05\%$ Tween 20 in PBS. Then block the membranes for about 1 h with 5% BSA/Tween 20 or overnight with 3% BSA/Tween 20. The membranes are then ready for a second round of incubation with primary antibody, wash, secondary antibodies, and ECL detection.

3.4. Flow-FISH for Mean Telomere Length Measurement

The average telomere length was measured using flow-FISH technology *(12)*. Cells were hybridized with a telomere-specific PNA probe ((AATCCC)$_3$-FITC, Applied Biosystems) and analyzed using flow cytometry.

3.4.1. Hybridization

1. Wash 2×10^5 cells two times with PBS in a 1.5-ml reaction tube and resuspend in 1 ml PBS + 0.1% BSA (*see* **Note 6**). Centrifuge the cells for 15 s at $15,000 \times g$, remove the supernatant (*see* **Note 7**), and resuspend the pellet in 200 µl hybridization mix (0.3 µg/ml PNA-Probe in 70% deionized formamide, 20 mM Tris, pH 7, 1% BSA), then store at −80°C (*see* **Note 8**). Incubate the samples for 5 min at 4°C, vortex, and denaturate in heat block for 10 min at 85°C (±0.5°C). (*see* **Note 9**) Incubate the samples for hybridization for 2 h (in the dark) (*see* **Note 10**). Add 1 ml wash I (70% formamide, 10 mM Tris, pH 7, 0.1% BSA, 0.1% Tween 20) to each sample, vortex, and centrifuge for 7 min at 18°C and $1720 \times g$.
2. Remove the supernatant leaving only 200 µl rest volume, and again add 1 ml wash I. Vortex the samples and centrifuge for 7 min at 18°C and $1720 \times g$. After removing the supernatant leaving only 200 µl rest volume, add 1 ml wash II (PBS + 0.1% BSA, 0.1% Tween 20), and vortex the samples and centrifuge for 7 min at 18°C and $740 \times g$.
3. Remove the supernatant leaving only 100 µl rest volume and add 300 µl propidiumiodide mix (0.05 µg/ml propidiumiodide, 10 µg/ml RNase in PBS + 0.1% BSA) and then vortex. Incubate the samples in the dark for 2 h at room temperature, then transfer into 5 ml polystyrene tube, and could be stored up to 2 days at 4°C.

3.4.2. FACS Acquisition

The cells are analyzed with FACScalibur Cytometer (Becton Dickinson, Heidelberg, Germany) (*see* **Note 11**). Using the DNA content (propidiumiodide in FL-3) against cell size (FSC) allows the detection of two cell populations: interphase cells with simple chromosome number (*2n*) and one population with doubled chromosome number (>4n). The telomere fluorescence (PNA-FITC in FL-1) is measured in a linear scale from the *2n* population region (*see* **Fig. 1**).

3.4.3. Analysis

1. The net telomere fluorescence is measured as the difference between the telomere fluorescence and autofluorescence, and the telomere length is expressed as *telomere fluorescence units* (TFU^{TRF}).
2. The TFU^{TRF} are based on calibration experiments using southern blotting ($n = 10$), which have shown that the mean telomere fluorescence is proportional to the TRF length ($r = 0.84$; $p < .0001$). The resulting slope in our FACS setting was $y = 0.0172x + 3.78$, indicating that a fluorescence intensity value of 0 corresponded

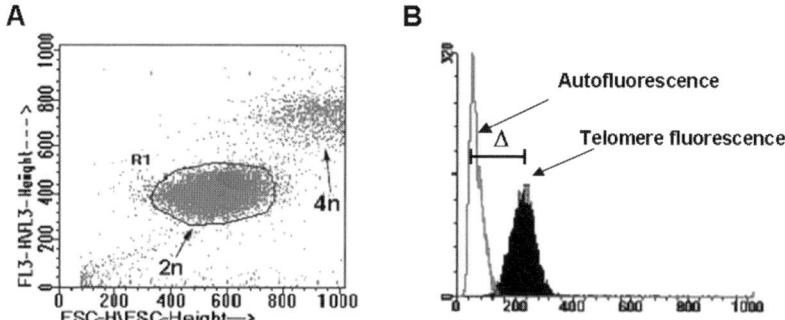

Fig. 1. Analysis of mean telomere length of peripheral blood cells using the flow cytometer (flow-FISH). (**A**) The fluorescence of gated *2n* DNA content is analyzed on a linear scale. (**B**) The telomere fluorescence is obtained by subtracting the mean fluorescence of the autofluorescence from the mean fluorescence obtained from cells hybridized with the telomere probe. The difference in fluorescence intensity (Δ) is expressed in *telomere fluorescence units* (TFU^{TRF}).

to 3.78 kbp in the Southern blot (*see* **Note 12**). This finding is similar to a previously reported calibration setup and may primarily reflect the fact that the telomere PNA probe does not bind to subtelomeric repeat sequences, which appear to be in the range of 2–4 kbp in length.

3. The intra-experimental variation of the flow-FISH analysis is about 2.1% (data not shown) (*see* **Note 13**). To compare different experiments, aliquots from mononuclear cells were run in duplicates with every experiment as controls (*see* **Note 14**). The mean fluorescence intensity of these controls was used as correction factor (k) for the measurement of other samples, and thereby one could decrease the inter-experimental variation from 14 to 3.2% (data not shown).

$$TFU^{TRF} \text{ (kb)} = \text{delta fluorescence}_{\text{sample}} \times k \times 0.0172 + 3.78$$

$$\text{delta fluorescence} = \text{fluorescence}_{\text{PNA-FITC}} - \text{autofluoresccence}$$

$$k = \frac{\text{delta fluorescence}_{\text{control sample calibration}}}{\text{delta fluorescence}_{\text{control sample acquisition}}}.$$

3.5. Cytogenetic Analyses and Q-FISH

3.5.1. Metaphase Preparation

1. After cell-culture treatment with 0.1 µg/ml colcemide for different time periods (depending on the cell type and proliferation capacity), harvest the cells, spin down, and incubate with 0.075 M KCl for 20 min at 37°C.

2. Incubate the cells with fixative methanol : acetic acid (3:1) for 15 min at room temperature. Repeat this step three times. Finally, spin down the cells and resuspend the pellet in 0.5 ml fixative. Drop the cells on the slides (previously cleaned with ethanol), and dry in water bath for 40 s at +50°C. The slides can be kept at room temperature prior to use for Q-FISH.
3. Signal-free ends (SFEs) are defined as chromosomal ends with no detectable telomere signal *(13)*. The frequency of SFE and chromosomal end-to-end fusions are determined by counting them in all metaphases and dividing by a number of metaphases.

3.5.2. Q-FISH

In principle, the procedure is very similar to flow-FISH, but the advantage is the possibility of determining the length of a single telomere on every chromosome (chromatid) end. Disadvantage is the relatively small number of analyzed cells limited to dividing cells. Q-FISH is performed as described previously *(14,15)*.

1. First rehydrate the slides for 15 min in PBS at room temperature, fix in 4% formaldehyde, and wash three times with PBS. After that incubate the slides with pepsin (550 U/ml), pH 2, for 10 min at 37°C.
2. Wash the slides two times with PBS and fix again in 4% formaldehyde. Wash three times with PBS and dehydrate the slides in serial of alcohol, for 5 min each. (70, 90, and 100%). Air-dry the slides and use for hybridization. Add $2 \times 10 \mu l$ hybridization mix on every slide and cover with coverslips (24×60).
3. Denature the slide for 2–3 min on 78–80°C in oven (*see* **Note 15**), put in a box, and hybridize for 2 h in humidity chamber on 37°C (*see* **Note 16**). Remove the coverslips carefully and wash the slides two times, for 15 min each, in washing buffer I and then three times with washing buffer II, for 10 min each, at room temperature.
4. Rehydrate the slides in serial of alcohol for 5 min each (70, 90, and 100%) and counterstained with DAPI. Dilute 1 mg/ml DAPI, 1:100 in PBS, and from that solution, prepare a new 1:25 dilution in Vectashield mounting media (Vector Laboratories). Add $2 \times 10 \mu l$ per slide and cover with coverslips (24×60).

3.5.3. Image Analysis

Before metaphase image analyses are done, daily fluctuations of the fluorescence lamp are excluded. Therefore, acquisition with calibration beads should be done first.

1. Dilute orange beads, $0.2 \mu M$ diameter (Molecular Probes, Eugene, OR), in protein solution (e.g., FCS/FBS) 1:25. Drop $2.5 \mu l$ bead/protein solution on coverslip (22×30) to make a thin homogenous smear on the glass.

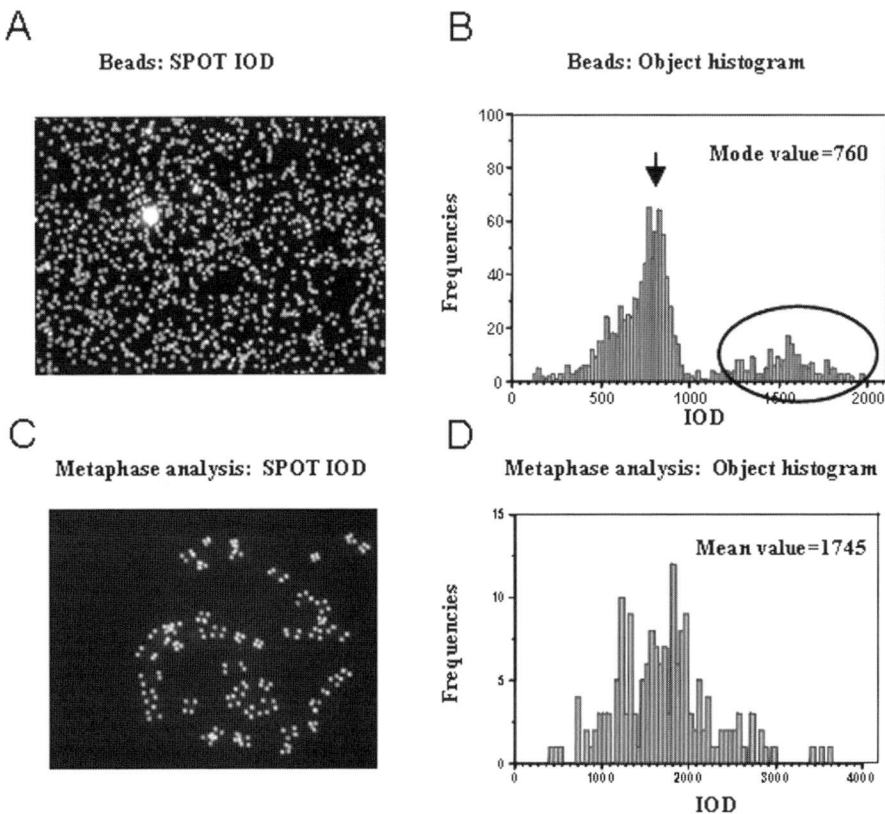

Fig. 2. Analysis of individual telomere length using quantitative fluorescence in situ hybridization (Q-FISH). (**A**) Beads immobilized on a glass slide are analyzed with the Spot integrated optical density (IOD) analysis of the TFL-TeloV1 software. The beads are shown as white spots. The green circles indicate identified spots by the Spot IOD analysis. (**B**) Using object histogram analysis, the mode of fluorescence intensity of beads is determined to exclude aggregated beads (circle). (**C**) Spot IOD analysis of telomeres in metaphase chromosomes from peripheral blood cells. (**D**) Telomere histogram of an individual metaphase. The integrated fluorescence intensity (IOD) for each telomere is calculated after correction for background, based on the values of the surrounding pixels, and after correction for image exposure time.

2. Air-dry the coverslips and keep in a covered box until use. Shortly before acquisition, drop 5 μl Vectashield mounting media (Vector Laboratories) on a glass slide and cover with bead-coated coverslip. These slides could be used for 3 days.
3. Digital images of beads and metaphase spreads are recorded with a digital camera (Sensys, Photometric) on an Axioplan II fluorescence microscope using the Vysis

workstation QUIPS (Vysis). The TFL-TELO software is used for analysis of telomere profiles. Telomere fluorescence intensity (TFI) values are expressed in arbitrary units [TFI = integrated optical density (IOD)].

4. First, capture 8–10 different images of the immobilized beads using the Cy-3 filter wheel by 200 ms exposure time. Then focus on the metaphases (at least 10) from target cells and capture the images (variable exposure time between 500 and 3000 ms) in their best focus plane.

5. Analyze the captured images with the TFL-TELO software starting with calibration images. First use the Spot IOD analysis (*see* **Fig. 2A**) followed by object histogram analysis (*see* **Fig. 2B**). Average mode value from 10 bead fields is defined as the bead calibration value (between 460 and 1000) *(16)*. Adapt every IOD value from metaphase object histograms (*see* **Fig. 2C** and **D**) to expose time and for bead calibration value. TFI values are expressed in arbitrary units (TFI = IOD).

4. Notes

1. After preparation of the hybridization mix, the one containing the PNA probes should be protected from light exposure all the time.

2. Addition of RNAase is important, because the propidium iodide staining used to distinguish the $2N$ population from populations with higher DNA content would also stain any RNA in the sample interfering with the wanted signal.

3. It is recommended to scan or photograph the membranes after ponso staining as proof of acceptable protein separation.

4. The duration of incubation with primary antibody can vary, we recommend performing preliminary experiments, as to optimize this incubation times.

5. The reuse of the primary antibody could be tried by storage at −25°C till needed (only with expensive antibodies).

6. The storage of samples at −80°C before the hybridization step seems not be limited. We successfully performed flow-FISH measurement on samples stored for more than 1 year.

7. Extra care should be taken while aspirating the supernatant as not to lose the pellet. In our experience, the use of vacuum suction machine was very reliable.

8. The volume of hybridization mix could be adjusted to the cell number available for flow-FISH measurement. For example, 1×10^5 cells in 100 μl and 5×10^5 in 500 μl. We do not recommend using less than 1×10^5 cells.

9. The heating oven should be preheated and temperature stabilized to 85°C, and this temperature should be closely monitored, as variation could strongly affect the results.

10. The room/space for the 2 h incubation should be air conditioned with no seasonal variation in atmospheric temperature as variation could strongly affect the results.

11. Calibration of the flow cytometer is performed weekly. However, this is only of minor importance for the flow-FISH inter-experiment variation, as internal control samples are used in every flow-FISH experiment.

12. Independent preliminary experiments using southern blotting for every laboratory is strongly recommended to validate the flow-FISH method.

13. The variation in flow-FISH results may be a major problem. It is mainly dependent on the following variables: (i) the heterogenous cell numbers in ratio to the telomeric probe, which would affect the fluorescence intensity; (ii) the inaccurate hybridization temperature we used a heating block instead of a water bath as it seemed to be more consistent; and (iii) the consistency of the amount of supernatant removed without disturbing the cell pellet during washing steps.

14. We used mononuclear cells mostly from healthy volunteers. The cells were separated using Ficoll, washed, counted, and stored as aliquots at −80°C. One aliquot is then used as internal control for every flow-FISH experiment.

15. Denaturation > 85°C may increase the telomere florescence, but this would result in loss the structure of metaphase and would not be able to define chromosomes as a cause of over denaturation.

16. In our experience, hybridization for 4 h did not make significant changes. We did not try to hybridize longer.

Acknowledgments

The authors thank Dr Stefan Zimmermann and Dr Milena Pantic for their kind advice and Dr Mike Harris for editing the manuscript. This work was supported by European Union (LSHC-CT-2004-502943), Sonderforschungsbereich 364 (Deutsche Forschungsgemeinschaft), Verein zur Leukaemieforschung, Freiburg, Germany.

References

1. Cech, T.R. (2004) Beginning to understand the end of the chromosome. *Cell*, **116**, 273–9.

2. Griffith, J.D., Comeau, L., Rosenfield, S., Stansel, R.M., Bianchi, A., Moss, H. and de Lange, T. (1999) Mammalian telomeres end in a large duplex loop. *Cell*, **97**, 503–14.

3. Cong, Y.S., Wright, W.E. and Shay, J.W. (2002) Human telomerase and its regulation. *Microbiol Mol Biol Rev*, **66**, 407–25, table of contents.

4. Zimmermann, S. and UM, M. (2005) Telomere dynamics in hematopoietic stem cells. *Curr Mol Med*, **5**, 179–85.

5. Shay, J.W. and Bacchetti, S. (1997) A survey of telomerase activity in human cancer. *Eur J Cancer*, **33**, 787–91.

6. Kelland, L. (2005) Overcoming the immortality of tumour cells by telomere and telomerase based cancer therapeutics–current status and future prospects. *Eur J Cancer*, **41**, 971–9.

7. Strahl, C. and Blackburn, E.H. (1994) The effects of nucleoside analogs on telomerase and telomeres in Tetrahymena. *Nucleic Acids Res*, **22**, 893–900.

8. Damm, K., Hemmann, U., Garin-Chesa, P., Hauel, N., Kauffmann, I., Priepke, H., Niestroj, C., Daiber, C., Enenkel, B., Guilliard, B., Lauritsch, I., Muller, E., Pascolo, E., Sauter, G., Pantic, M., Martens, U.M., Wenz, C., Lingner, J., Kraut, N., Rettig, W.J. and Schnapp, A. (2001) A highly selective telomerase inhibitor limiting human cancer cell proliferation. *EMBO J*, **20**, 6958–68.
9. Pascolo, E., Wenz, C., Lingner, J., Hauel, N., Priepke, H., Kauffmann, I., Garin-Chesa, P., Rettig, W.J., Damm, K. and Schnapp, A. (2002) Mechanism of human telomerase inhibition by BIBR1532, a synthetic, non-nucleosidic drug candidate. *J Biol Chem*, **277**, 15566–72.
10. El-Daly, H., Kull, M., Zimmermann, S., Pantic, M., Waller, C.F. and Martens, U.M. (2005) Selective cytotoxicity and telomere damage in leukemia cells using the telomerase inhibitor BIBR1532. *Blood*, **105**, 1742–9.
11. Rufer, N., Dragowska, W., Thornbury, G., Roosnek, E. and Lansdorp, P.M. (1998) Telomere length dynamics in human lymphocyte subpopulations measured by flow cytometry. *Nat Biotechnol*, **16**, 743–7.
12. Blasco, M.A., Lee, H.W., Hande, M.P., Samper, E., Lansdorp, P.M., DePinho, R.A. and Greider, C.W. (1997) Telomere shortening and tumor formation by mouse cells lacking telomerase RNA. *Cell*, **91**, 25–34.
13. Martens, U.M., Chavez, E.A., Poon, S.S., Schmoor, C. and Lansdorp, P.M. (2000) Accumulation of short telomeres in human fibroblasts prior to replicative senescence. *Exp Cell Res*, **256**, 291–9.
14. Martens, U.M., Zijlmans, J.M., Poon, S.S., Dragowska, W., Yui, J., Chavez, E.A., Ward, R.K. and Lansdorp, P.M. (1998) Short telomeres on human chromosome 17p. *Nat Genet*, **18**, 76–80.
15. Poon, S.S., Martens, U.M., Ward, R.K. and Lansdorp, P.M. (1999) Telomere length measurements using digital fluorescence microscopy. *Cytometry*, **36**, 267–78.
16. Zijlmans, J.M., Martens, U.M., Poon, S.S., Raap, A.K., Tanke, H.J., Ward, R.K. and Lansdorp, P.M. (1997) Telomeres in the mouse have large inter-chromosomal variations in the number of T2AG3 repeats. *Proc Natl Acad Sci USA*, **94**, 7423–8.

7

Uses of Telomerase Peptides in Anti-Tumor Immune Therapy

He Li, Indzi Katik, and Jun-Ping Liu

Summary

Human telomerase reverse transcriptase (hTERT) represents a universal tumor-associated antigen to activate specific immune response in cancer immune therapy. Peptides derived from hTERT are presented by major histocompatibility complex (MHC) class I alleles to T lymphocytes, and CD8+ cytotoxic T lymphocytes (CTLs) specific for the hTERT-derived antigenic epitopes lyse hTERT-positive tumors from multiple histologies. These findings identify hTERT as an important tumor antigen widely applicable for anti-cancer immunotherapeutic strategies. The hTERT antigen-specific immunotherapy involves both active vaccination and adoptive immunotherapy approaches. Most importantly, the anti-tumor immune responses have been observed in the absence of toxicity, underlying the ongoing endeavors to develop immunotherapy directed against hTERT antigen. This chapter discusses most promising results and the approaches for investigation to target hTERT peptides as tumor antigens.

Key Words: Telomerase; hTERT; peptide; lymphocyte; cancer immunotherapy.

1. Introduction

Cancer immunotherapy depends on the identification of tumor antigens that can induce a clinically efficient anti-tumor immune response. Specific tumor-associated antigens (TAAs) were initially demonstrated by identifying patient cytotoxic T cells that recognize antigenic peptides in early 1990s *(1,2)*. TAAs have since been identified from tumors of multiple histologies, including

From: *Methods in Molecular Biology, vol. 405: Telomerase Inhibition*
Edited by: L. G. Andrews and T. O. Tollefsbol © Humana Press Inc., Totowa, NJ

melanoma, breast cancer, and prostate cancer *(3–8)*. Many studies have been carried out to explore the insufficiency of tumor-specific responses to the existing TAAs. Evidence suggests that although high-frequency T lymphocytes specific for certain TAAs have been detected, these T cells may be functionally inactive in vivo failing to induce meaningful clinical responses *(9,10)*. The immune evasion is thought to be because of multiple defects including inefficient antigen processing, lack of cytokine production, and impairment of antigen-specific cytolytic function *(11,12)*.

Most tumor antigens are cytoplasmic self-antigens that are either overexpressed or selectively expressed by tumor cells. Even though the antigens are immunologically tolerated during tumor development, they become capable of triggering effector T-cell responses against tumors after ex vivo re-stimulation with cytokines and optimal presentation by antigen-presenting cells (APCs). These findings have triggered a new light in clinical efforts to target TAAs, vaccination, and use of adoptive T-cell therapy *(3)*. CD8+ cytotoxic T lymphocytes (CTLs) are the principal effector cells of antigen-specific anti-tumor immune responses. The specificity of TAA-triggered anti-tumor immune response is ensured by the binding of TAA receptors on the surface of CTL to 9–10 amino acid TAA peptides displayed by major histocompatibility complex (MHC) class I molecules on the surface of tumor cells. This engagement between antigenic peptide/MHC class I and its receptors results in a conventional T-cell response consisting of cytokine secretion and target cell lysis (*see* **Fig. 1**).

Although analyses of tumor-derived T-cell clones and antibodies from patients reveal antigens specifically co-existing with particular tumors *(13,14)*, the method of matching antigenic peptide epitopes with MHC class I binding motifs from genes known to be selectively expressed in tumors has proven to be effective in identifying novel antigens (reviewed in **ref. 15**). Ideal TAA is expected to be expressed by the vast majority of human cancers, be processed by tumor cells for presentation by MHC molecules, be recognized by the T-cell repertoire in an MHC-restricted mode, and most importantly, stimulate expansion of the CTL precursors bearing specific receptors. One such TAA that bears all these features is the catalytic subunit of telomerase [human telomerase reverse transcriptase (hTERT)] *(3)*. As hTERT is immunologically tolerant or incompetent to elicit effective immunity under progressive tumor burden *(9,10)*, it could not be defined as a TAA from patient immunoreactivity, but its antigenic epitopes have been deduced from hTERT primary sequence and characterized by methods of "reverse immunology" *(16)*.

T Cell Targets hTERT Positive Tumor Cell

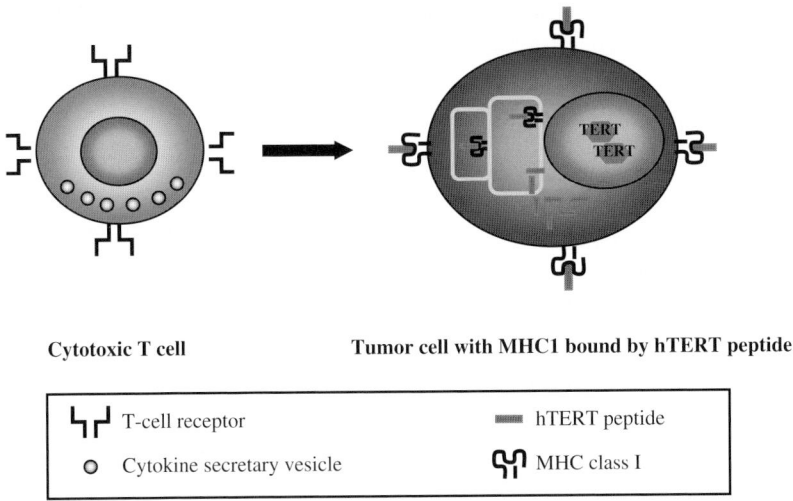

Cytotoxic T cell   **Tumor cell with MHC1 bound by hTERT peptide**

T-cell receptor	▬▬▬ hTERT peptide
○ Cytokine secretary vesicle	MHC class I

Fig. 1. Schematic illustration of human telomerase reverse transcriptase (hTERT) antigenic peptide presented by major histocompatibility complex (MHC) class I on cancer cell surface. The peptides (thin bar) of hTERT are produced in cytoplasm by proteasome degradation of hTERT proteins that usually operate in the nucleus. Depredated hTERT peptides are transported by transporter associated with antigen processing into endoplasmic reticulum where the peptides are loaded onto MHC class I molecules and further transported through secretary pathways and presented on cell surface. Cytotoxic T lymphocyte with specific receptors recognizes and targets the cancer cell by engagement with the hTERT peptide antigen and MHC class I.

1.1. hTERT Peptides as Tumor Antigens

As hTERT is identified as a tumor antigen by deduction, it remains to be fully established that hTERT can trigger a natural in vivo T-cell response that may be functionally inactivated during tumor progression. Evidence from both human and murine studies shows that CTLs recognize peptides derived from hTERT or murine TERT (mTERT) and eliminate TERT-positive tumor cells of multiple histologies *(16)*. With the ability of naïve T cells to expand specific CTLs after multiple re-stimulations by hTERT antigens ex vivo (*see* **Fig. 2**), the T-cell repertoire against hTERT is apparently intact in normal

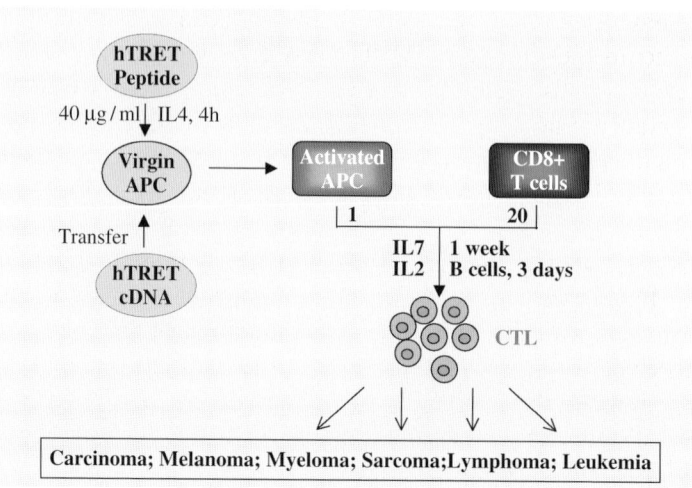

Fig. 2. Schematic procedure for generation of human telomerase reverse transcriptase (hTERT) peptide-specific cytotoxic T lymphocytes (CTLs) to target multiple types of cancer cells positive in telomerase activity. Unstimulated antigen-presenting cells (APC) are pulsed with synthetic hTERT peptides in the presence of interleukin-4 (IL-4) or infected with defective recombinant virus encoding hTERT or hTERT peptides. Activated APCs are co-cultured with naïve CD8-positive CTLs in the presence of IL-7, IL-2, and B lymphocytes for different periods of time as indicated. Activated CTLs are tested in cultured cancer cells and in vivo.

donors and cancer patients *(17–19)*. Thus, hTERT may represent as a universal TAA expressed in most cancers *(20,21)*. Using hTERT as an immune target may minimize immune evasion induced by antigen loss, as hTERT is critically required for tumorigenesis *(22–24*; also see recent review in **ref. 25**). These together suggest that hTERT is an attractive target for novel immunotherapy against various cancers.

Examining the primary sequence of hTERT for peptides that could potentially bind to MHC class I molecules by a peptide motif scoring system has suggested multiple antigenic peptides. The first characterized hTERT-derived peptide is hTERT I540–548 peptide that binds to the most common human leukocyte antigen (HLA) subtype—HLA-A2 that is an MHC class I allele expressed by nearly 50% of cancer patients. The hTERT-derived peptide I540 (ILAKFLHWL) is identified within the middle of hTERT protein sequence, approximately 70 amino acids to the amino terminus of the first reverse transcriptase motif. Vonderheide and colleagues *(3)* have suggested that the telomerase-derived antigenic peptide(s) can be naturally processed and presented in the groove of MHC class I on the surface of tumor cells (*see* **Fig. 1**). Currently identified hTERT CTL epitopes with HLA restriction would potentially render up to 75% of cancer patients eligible for hTERT-specific immunotherapy, and the total number of hTERT candidate antigens is till increasing by the method of epitope deduction from bindings to HLA subtypes.

The hypothesis that telomerase may be a clinically important tumor rejection antigen has been tested by immunization of mice with dendritic cells (DCs) being transfected with mTERT. The novel findings of immune surveillance against cancer cells without toxicity opened the door for the consideration of preventative immunotherapy for healthy individuals considered at high risk for cancer with genetic factors and medical history *(15)*. With the HLA-A2-restricted hTERT I540 peptide, a phase I clinical trial has been performed to evaluate clinical and immunological impact of vaccinating advanced cancer patients *(18,26)*. In a mouse model, immunization with mTERT mRNA-transduced DCs has demonstrated the generation of TERT-specific protective immunity without the development of autoimmunity against TERT-expressing cells *(27)*. As telomerase is expressed widely in normal mouse tissues, different from the restricted expression in humans *(28)*, immune tolerance may limit responses to TERT in mouse more than that in human *(27)*. Successful anti-tumor vaccine depends on a pre-existing T-cell pool capable of recognizing epitopes from the relevant antigen that can be activated in vivo. Thus, whether a low-frequency population, in contrast to an expanded but functionally inactivated precursor population, would facilitate clinical attempts to induce anti-tumor immunity requires further investigation.

A number of synthetic peptides derived from hTERT have been investigated that induce specific CTLs. Peptides hTERT I540 (ILAKFLHWL), hTERT R865 (RLVDDFLLV) *(3)*, hTERT 572 (RLFFYRKSV), and hTERT 988 (DLQVNSLQTV) *(5)* react in a manner dependent on HLA-A2 *(29)*. Peptide K973 (KLFGVLRLK) acts in an HLA-A3-restricted manner, with HLA-A3 expressed by 15–25% of patients *(30)*. Peptides TEL324 (VYAETKHFL) and TEL461 (VYGFVRACL) bind in an HLA-A24-restricted manner; HLA-A24 is the most common HLA allele among Japanese *(31)* and also frequently present in persons of European descent *(32)*. Observably, each of these peptides has unique binding sites encoded by the MHC alleles in human genome, and with increased expression of HLA subtypes on T cells, each peptide may be processed by particular MHC restriction molecules and bind to particular T-cell receptors to induce CTL production *(16)*.

1.2. hTERT-Induced Anti-Tumor Cytotoxicity Versus Immune Tolerance and Autoimmunity

Studies undertaken with the hTERT peptides have shown that HLA-restricted CTLs against hTERT are equally induced ex vivo from cancer patients and healthy individuals and efficiently destroy tumor cell lines and primary tumors from a wide range of histologies *(33)*. Peptide/MHC tetramer assays are used to specify whether there are hTERT-specific CTL responses measurable at baseline in cancer patients. In uncultured peripheral blood, neither healthy volunteers nor cancer patients exhibit an expanded pool of specific CTLs for the HLA-restricted hTERT epitope suggesting that hTERT is tolerated at baseline by the immune system even in the setting of active neoplasia *(30)*. However, specific CD8+ CTL response against hTERT peptide is induced after at least three rounds of ex vivo peptide stimulation, evaluated by a flow cytometric MHC tetramer assay. These findings that CTLs against hTERT are consistently induced ex vivo despite basal immune ignorance to the existent hTERT TAA suggest that small populations of hTERT-specific T-cell precursors have escaped irreversible elimination from central immune tolerance mechanism and can be stimulated by hTERT TAAs for immunity against telomerase-positive tumors.

A crucial concern of the hTERT-derived antigen immunotherapy approach is the possibility of cytolysis of some normal cell types in which telomerase can be detected. These cells include hematopoietic stem cells and progenitors, germinal center cells, basal keratinocytes, gonadal cells, and certain proliferating epithelial cells *(34–38)*. Therefore, whether hTERT I540 peptide-specific CTLs might lyse normal cells has been examined in human peripheral blood CD34 cells positive in telomerase activity. No lysis by I540-specific CTLs is observed on CD34-enriched cells, suggesting that hTERT does not function as an auto-antigen on hematopoietic precursor cells, probably because of low levels of hTERT expression and lack of antigen processing in CD34-enriched cells *(3,17,33)*. Consistently, the findings that I540-specific CTLs does not lyse stem cells and progenitors and that the level of hTERT expression is not rate limiting for CTL activation suggest an inability of the cells to process this antigen and present it in MHC class I *(3)*. Thus, hTERT appears to be a poor auto-antigen of hematopoietic stem cells, activated T lymphocytes, and other normal cells that have telomerase activity.

As dominant peptides abundant on cell surface with high affinity for HLA class I may cause clonal deletion of their T-cell precursors, some peptides with low HLA class I affinity and weak presentation have been investigated to optimize by sequence modification to increase their binding to MHC for immunity to subject them to be more immunogenic *(5)*. These peptides are often referred as cryptic peptides with the mutated peptides as heteroclitic peptide analogues. Optimized cryptic peptides homologous to TERT have been shown to induce efficient anti-tumor T-cell cytotoxic immunity without autoimmunity in vivo in HLA-A*0201 transgenic mice, healthy blood donors, and prostate cancer patients *(19)*. Thus, it has been suggested that the T-cell repertoire specific for cryptic peptides has avoided being clonally deleted in the tymus, and thus cryptic peptides are better targets than dominant peptides for cancer vaccination after having turned them to be immunogenic *(19)*. A most attractive cryptic hTERT peptide may be Y572, for whose optimization ODN-CpG is used as adjuvant, that elicits effective anti-tumor T-cell cytotoxic immunity in vivo *(19)*. Thus, the clinical trials demonstrate that hTERT cryptic peptides are promising candidates for cancer immunotherapy.

1.3. Specificity Counts

The methods of cytotoxicity analysis have been performed to evaluate the cytotoxic effector function of hTERT CTL lines. Multiple studies have demonstrated CTL specificity for hTERT by comparing the tumor cells with controls negative for hTERT. To test whether hTERT peptide-derived specific CTL can recognize particular antigen (peptide specificity) in a broad range of tumors, their cytotoxicity is evaluated against control peptides not from hTERT. For I540 peptide, the CTLs demonstrate peptide-specific cytotoxicity of cell targets pulsed with the I540 peptide compared with targets pulsed with irrelevant peptides, such as F271 MAGE-3 or I476 RT-pol or 2-microglobulin alone *(33)*. Carried out with hTERT-negative cell line infected either with a retroviral vector encoding the full sequence of hTERT or with a virus specifying only a drug-resistance marker, it has been shown that hTERT-specific CTLs fail to lyse the control cell line but lyse the hTERT-expressing cell line effectively *(33)*. As both cell lines expressed similar levels of particular HLA allele and nearly equal amounts of surface MHC class I (determined by flow cytometry), the hTERT antigenic peptide specificity of CTLs is demonstrated *(3,17,33)*.

Tumor specificity of patient-derived hTERT-specific CTLs has been further evaluated by MHC subtyping in a panel of telomerase-positive tumor cell lines, which differ in the presence or absence of particular HLA allele of interest. Cytotoxicity is only observed in cell lines derived from patients bearing the HLA allele of interest but not in HLA-negative tumor cell lines. MHC-restricted cytotoxicity has also been challenged by use of anti-HLA class I monoclonal antibodies (mAbs) and control anti-CD20 mAbs. Consistent with MHC restriction, peptide-specific lysis is blocked by addition of anti-HLA antibodies, whereas control anti-CD20 mAbs does not inhibit peptide-specific lysis of tumor targets *(6)*, suggesting that hTERT is recognized as an antigen by the CTLs in an HLA-restricted manner. To attest whether the cytotoxicity of hTERT peptide-specific CTL clones may attack cancerous cells by specific binding to antigenic epitopes of endogenously processed hTERT, cold peptide competitive inhibition experiments have been performed. In this experiment, cytotoxicity is inhibited by the addition of hTERT peptide-loaded autologous cells, suggesting that hTERT has been naturally processed in cancerous cells and that the derived peptides are expressed on these cells and recognized by hTERT-specific CTLs in the context of HLA *(32)*.

Consistent with the concept that the specificity of antigen-induced cytotoxicity is regulated by APC, in vivo T-cell priming by DCs has been observed *(39)*. In addition, the induction of CTL responses and tumor immunity against unrelated tumors has been determined independently *(27)*. The mean stimulation indices range from about 10 to 50, depending on the formulation (mature vs. semi-mature), type of subject (healthy individual vs. cancer patient), and route of administration of hTERT RNA transfected DCs *(39–41)*. Two small trials in metastatic melanoma suggest that mature monocyte-derived DCs are more effective at inducing immune responses than immature DCs *(42,43)*. Although immune responses have been observed previously with immature or semi-mature DCs, particularly if injected intranodally *(41,44)*, fully immature DCs used in two other studies fail to prime immune response *(43,45)*. Thus, DCs at different stages of maturity possess different capacities to process and present hTERT TAA. These studies together suggest that in addition to the expression levels of hTERT and its availability, cells with different maturities in antigen processing and presentation regulate not only CTL production but also specific interactions with hTERT antigenic peptides.

Finally, hTERT peptide-specific CTLs have been shown specific cytotoxicity against target tumor cells from a wide variety of histological origins including carcinoma, malignant melanoma, multiple myeloma, sarcoma, non-Hodgkin's lymphoma, leukemia, and Epstein–Barr virus (EBV)-transformed B cells *(3,8,46)*. Although differences exist in study design, TAA structure, source of precursor cells, maturation stage, dose, and route of administration, consistent findings from in vitro cell cultures and in vivo animal models and clinical trials converge on a consensus of supporting more basic and clinical researches to target hTERT as a tumor antigen with broad therapeutic potentials. The following sections detail the procedures used in studies of hTERT peptides in various models, serving as a framework for further investigations of feasibility of cancer immunotherapy and its opportunities and challenges.

2. Materials

2.1. Patient Samples and Healthy Donors (3,32)

1. Human peripheral blood mononuclear cells (PBMCs) were obtained by phlebotomy or leukapheresis followed by Ficoll-density centrifugation.
2. Primary tumor cells were obtained following biopsies.
3. CD34-enriched peripheral blood cells were obtained from lung cancer patients and a patient with sarcoma scheduled for autologous stem cell transplantation and purified by cell sorting.
4. Acute myelogenous leukemia (AML) and non-Hodgin lymphoma (NHL) cells were obtained from discarded clinical material.
5. HLA-A2+ individuals were selected by fluorescence-activated cell sorter (FACS) screening by using mAB BB7.2 (*see* **Note 1**).

2.2. Cell Lines

Most cell lines were obtained from American Type Culture Collection (Manassas, VA) including TAP-T2 cells (antigen transporting deficient HLA-A2.1-positive human T2 cells), the transformed kidney cell line 293, the Calu-1 lung carcinoma cell line, the multiple myeloma cell lines U266, IM9, and human serum (HS)-Sultan, and EBV-transformed B cell line SKW6.4. The 36M ovarian carcinoma cell line was generated from patient ascites. The K029 and K017 malignant melanoma cell lines were generated from biopsies of HLA-A*0201-positive and HLA-A*-negative patients with melanoma. B-lymphoblastoid cell lines (B-LCLs) were established by the transformation of peripheral blood B lymphocytes using EBV and EBV-transformed B cell line.

2.3. Cytokines

Interleukin-2 (IL-2) (Chiron, Emeryville, CA), IL-4 (Immunex, Seattle, WA), IL-7 (R&D Systems, Oxon, UK), IL-10 (R&D Systems), granulocyte-macrophage colony-stimulating factor (GM-CSF) (Genzyme, Cambridge, MA), and tumour necrosis factor-α (TNF-α) (Genyzme).

2.4. Media, Antibodies, and Reagents

1. Cell-culture media: Iscove's MDM (BioWhittaker, Verviers, Belgium), RPMI 1640 (for culturing PBMCs and hTERT-transfected cell lines), and RPMI 5 (for culturing cells required for in vitro induction of CTLs) were from Gibco/BRL, Carlsbad, CA, USA. β2-microglobulin. Yssel's medium (for in vitro induction of CTLs).
2. Antibodies: antibodies against CD8+-fluorescein isothiocyanate (FITC) (Immunotech, Marseilles, France), CD4-PerCP (Becton-Dickinson, San Jose, CA), and CD14-PerCP (Becton-Dickinson). For phynotypic analysis, mABs against

CD45RA-allophycocyanin, CD45RO-APC, CD28-APC, CD27-FITC (PharMingen, San Diego, CA), anti-CCR7-FITC (R&D Systems), and CD8-APC or CD8-FITC (Immunotech). mAB against interferon-γ (IFN-γ) (Mabtech, Nacka, Sweden). Anti-CD8β mAB and microbead-conjugated anti-mouse immunoglobulin G (IgG) antibodies (Miltenyi Biotech, Bergisch Gladbach, Germany). Polystyrene beads coated with an anti-CD8 mAB (DYNAL, Oslo, Norway). FITC-conjugated anti-HLA mABs against various HLA alleles (One Lambda, Canoga Park, CA). Fluoresceinated F(ab')$_2$ rabbit anti-mouse IgG (Serotec, Oxford, UK).
3. Human β2-microglobulin (Sigma-Aldrich, St. Louis, MO).
4. Phytohemagglutinin (PHA) (Murex Biotech, Dartfort, UK).
5. Streptavidin-alkaline phosphatase (Mabtech).
6. 5-bromo-4-chloro-3-indolylphosphate and introblue tetrazolium color development substrate (Promega, Madison, WI).
7. Brefeldin A (Sigma-Aldrich).
8. HS (ICN Biomedicals) phycoerythrin.
9. Transferrin (Roche Diagnostics Australia Pty Ltd, Castle Hill, NSW, Australia).
10. Insulin (Sigma-Aldrich, Castle Hill, NSW, Australia)
11. Glutamine and gentamicin (Gibco/BRL)
12. Hygromycin B (Sigma, St. Louis, MO).
13. Buffy coats from normal donors were purchased from the San Diego Blood Bank (San Diego, CA). Caltag Fix and Perm Kit (Caltag, Burlingame, CA) for intracellular staining of cytokines.

2.5. Peptides

For clinical administration, hTERT I540 peptide was purchased from Multiple Peptide Systems (San Diego, CA) as Good Manufacturing Practice (GMP)-grade lyophilized powder. Otherwise, peptides were commercially synthesized and purified (>90%) by reversed-phase high-performance liquid chromatography from several sources including Macromolecular Resources (Fort Collins, CO) and Sigma Chemical Co. Genosys Biotechnologies (The Woodlands, TX).

1. HLA-A2-restricted hTERT peptides: I540 (ILAKFLHWL), R865 (RLVDDFLLV), R572 (RLFFYRKSW), and R988 (DLQVNSLQTV); control peptides: I476 (ILKEPVHGV) from RT-pol of HIV, F271 (FLWGPRALV) from MAGE-3, G58 (GILGFVFTL) from the matrix protein of influenza A, and L11 (LLFGYPVYV) from the tax protein of HTLV-1.
2. HLA-A3-restricted hTERT peptides: K973 (KLFGVLRLK). Control peptides: I265 (ILRGSVAHK) from the nucleoprotein of Influenza NP and I476 (ILKEPVHGV) from HIV RT-pol gene, both were used as positive controls *(47–49)*. Peptide I540 also used as a negative control here for HLA-A3 restriction binding.
3. HLA-A24-restricted hTERT peptides: V324 (VYAETKHFL) and V461 (VYGFVRACL); control peptides: I195 (IMPKAGLLI) from MAGE-3 *(50)* and

R583 (RYLKDQQLL) from human immunodeficiency virus type 1 (HIV-1)gp41 were used as positive controls for HLA-A24 binding and characterization of hTERT-specific HLA-A24-restricted CTLs.

2.6. Animal Models

1. In vivo immunogenicity was assessed in HHD (H-2 D^b and β2-microglobulin double knockout and the α1 and α2 of HLA-A2.1 and α3 of H-2 D^b chimeric transgene linked with β2-microglobulin) mice *(51)*. These mice are H-2 $D^{b-/-}$ and β2-microglobulin$^{-/-}$ and express chimeric MHC class I molecule with the α1 and α2 domains of HLA-A2.1 and α3 domain of H-2 Dβ$^{-/-}$, to preserve the interaction with murine CD8, and covalently linked with the human β2-microglobulin light chain *(29)*.
2. Humanized HLA-DR4 transgenic mice (HLA-DRB1*0401), which are murine class II deficient and transduced with human CD4 molecule, were generated by Dr Grete Sonderstrup in the Department of Microbiology and Immunology of Stanford University *(52–54)*.
3. C57BL/6 (H-2^b) mice were vaccinated with TERT RNA-transfected DCs to measure anti-tumor immunity in vivo *(27)*.

2.7. Other Materials

1. ImmunoSpot plates (Cellular Technology, Cleveland, OH, or Millipore, Molsheim, France) for Enzyme-Linked Immunospot (ELISPOT) assay.
2. ELISPOT reader (Carl Zeiss, Berlin, Germany)
3. FACS Calibur flow cytometer and Cell Quest software (BD, North Ryde, NSW, Australia)
4. 5% CO_2 incubator for cell culture.

3. Methods
3.1. Identification and Analysis of MHC-Restricted hTERT Peptides

1. The amino acid sequence of hTERT (locus AF015950) was analyzed for sequences containing known binding motifs for MHC molecules *(55)*.
2. Peptides were identified by reverse genetics based on the canonical anchor residues for HLA-A2 *(56)* and the software of the Bioinformatics and Molecular Analysis Section (National Institutes of Health, Washington DC) available at http://www-bimas.cit.nih.gov/molbio/hla_bind/index.html, which ranks 9-mer peptides on a predicted half-time dissociation coefficient from HLA class I molecules *(57)*.

3.2. HLA Typing

1. Indirect flow cytometric analysis was performed on all PBMC samples to identify subjects expressing HLA-A2 or -A3 or -A24. Murine mABs specific for these HLA alleles were used. Fluoresceinated F(ab')$_2$ rabbit anti-mouse IgG was used as a secondary detection reagent.

2. HLA serotyping was performed by a microlymphocyte cytotoxicity assay with qualified antisera. According to serologic typing results, HLA class I alleles were amplified by polymerase chain reaction (PCR) with group-specific primers, then typed at nucleotide sequence levels. For some HLA such as HLA-A24, expression in some leukemia cells was examined by flow cytometry with a FITC-conjugated anti-HLA-A24 mAB and FITC-conjugated mouse IgG as control.

3. The samples were analyzed on a FACS Calibur flow cytometer running Cell Quest software.

3.3. HLA-A2.1 Binding/Stabilization Assay

1. Although the immunogenicity of MHC class I-restricted peptides to some degree reflects their binding and stabilizing capacity for MHC class I molecules, a direct proof of the strength of interaction between the two hTERT peptide and the HLA-A2.1 molecule was achieved by measuring the relative avidity of peptide to HLA-A*0201 with the HLA-A2 binding/stabilization assay *(58)*.

2. Antigen-transporting deficient (TAP)-HLA-A2.1-positive human T2 cells were incubated overnight at 37 °C in RPMI 1640 medium supplemented with human β2-microglobulin (100 ng/ml) (Sigma) in the absence (negative control) or presence of peptide to be tested or reference peptide of HIV reverse transcriptase at various final peptide concentrations (0.1–100 μM).

3. Cells were incubated with Brefeldin A (0.5 μg/ml) for 1 h and subsequently stained with a saturating concentration of mAB BB7.2 for 30 min at 4 °C, followed by washing and a second incubation with a goat antibody to mouse Ig F(ab′)$_2$ conjugated to FITC (Caltag, South San Francisco, CA).

4. Cells were then washed, fixed with 1% paraformaldehyde, and analyzed in a FACS Calibur cytofluorimeter (Becton-Dickinson). The mean fluorescence intensity of each concentration minus that of cells without peptide was used as an estimate of peptide binding and results expressed as values of relative avidity, which is the ratio of the concentration of test peptide necessary to reach 20% of the maximal binding by the reference peptide over that of the reference peptide, so that the lower the value the stronger the binding (*see* **Note 2**).

3.4. In Vivo Evaluation of Baseline CTL Responses to the hTERT Peptide

To evaluate the extent to which cancer patients and normal donors respond in vivo to the hTERT peptides, analysis of MHC class I tetramers on PBMC from patients and healthy donors was performed as below.

3.4.1. Analysis of MHC Class I Tetramers

1. Soluble HLA-A2 tetramers were prepared with immunizing peptides and β2-microglobulin as described *(59)*, conjugated to phycoerythrin, and validated using

peptide-specific CTLs as described *(33)*. Control tetramer was made with the HLA-A2-binding peptide LII (LL-FGYPVYV) *(60)* from HTLV-1 tax and validated using LII-specific clones *(33)*.

2. Cells were incubated with tetramers and with mABs against CD8+-FITC (Immunotech), CD4-PerCP (Becton-Dickinson), and CD14-PerCP (Becton-Dickinson) for 30 min at room temperature.

3. For phynotypic analysis, mABs were CD45RA-APC, CD45RO-APC, CD28-APC, CD27-FITC (PharMingen), anti-CCR7-FITC (R&D Systems), and CD8-APC or CD8-FITC (Immunotech).

3.4.2. ELISPOT Assay

1. Uncultured in vitro-stimulated PBMCs at 2.5×10^4 cells/well were added to ImmunoSpot plates (Cellular Technology) pre-coated with 10 ug/ml anti-IFN-γ mAB (Mabtech) in the presence or absence of 5 ug/ml peptide overnight at 37 °C.

2. After washing, wells were then incubated with 1 ug/ml biotin-conjugated anti-IFN-γ mAB against IFN-γ followed by streptavidin-alkaline phosphatase (Mabtech). Purified anti-CD3 mAB was used as a positive control.

3. Spots were developed with 5-bromo-4-chloro-3-indolyl-phosphate and introblue tetrazolium color development substrate (Promega).

4. Spots were counted using a Prior ProScan analyzer and Image Pro Plus software (Hitech Instruments, Edgemont, PA). Specific spots were calculated as the number of spots with peptide subtracts the number of spots without peptide.

3.5. In Vitro Generation of hTERT Peptide-Specific CTL Clones (see Fig 2)

3.5.1. Generation of DCs from PBMCs

1. PBMCs were depleted of T, B, and natural killer (NK) cells by magnetic bead depletion *(61,62)*. The remaining cell fraction (1.3×10^6 cells/ml, >80% CD14+) was stimulated with GM-CSF (50 ng/ml, Genzyme) and IL-4 (10 ng/ml, Immunex) in Iscove's MDM (Gibco/BRL) supplemented with 5% human AB serum, 50 ug/ml transferring (Roche), 5 ug/ml insulin (Sigma Chemical Co.), 2 mM glutamine (Gibco/BRL), and 15 ug/ml gentamycin (Gibco/BRL) at 37 °C in 5% CO_2 to generate DCs.

2. To determine maximum expansion, DCs were cultured up to 30 days with GM-CSF and IL-4. Cytokines were added at the beginning of culture and every third day thereafter. For functional analysis, DCs were matured on day 6 for 48 h with either TNF-α (30 ng/ml, Genyzme) or t-CD40L before use as APCs in allogenic mixed lymphocyte reactions (allo-MLRs).

3. On day 8 or 9, cells were harvested and used as monocyte-derived DCs for antigen presentation. To determine DCs, the associated antigens such as CD1a, CD80, CD83, CD86, and HLA classes I and II were measured by FACS *(63)*.

3.5.2. Generation of hTERT-Specific CTLs

3.5.2.1. ISOLATION OF CD8β-POSITIVE CTL PRECURSOR

1. PBMCs from healthy donors or patients were isolated from certain types of HLA allele-positive buffy coat by density gradient centrifugation using lymphoprep (Nycomed, Oslo, Norway).
2. Isolation of resting CD8β-positive CTL precursors from total PBMC was performed by positive selection on an automated magnetic sorting device (autoMACS, Miltenyi Biotech). For this purpose, total PBMCs were stained with anti-CD8β mAB and microbead-conjugated anti-mouse IgG antibodies (Miltenyi Biotech), followed by autoMACS sorting.
3. The CD8β-negative PBMC fraction was used to generate mature monocyte-derived DCs as described above in **Section 3.5.1.** Plastic-adherent PBMCs (8×10^6/ml in Iscove's supplemented with 1% HS, ICN Biomedicals, for 45–60 min at 37 °C) were washed with PBS and cultured for 6–7 days in Iscove's supplemented with 8% FCS, 1000 U/ml IL-4 (Sanquin, Amsterdam, The Netherlands), and 100 ng/ml GM-CSF (Schering-Plough, Kenilworth, NJ).
4. Cytokines were refreshed at days 3 and 6, and DC maturation was induced by an additional 2-day culture in the presence of monocyte-conditioned medium, generated as described *(64)*.

3.5.2.2. MATURATION OF DCs

1. PBMCs (50×10^6) were seeded on a human Ig-coated (30 ug/ml in PBS for 1 h at room temperature) 10-cm bacteriological Petri dish and incubated for 30 min at 37 °C in Iscove's supplemented with 2% HS.
2. Supplement (monocyte-conditioned medium) was harvested, filtered, and used at 25–50% v/v for DC maturation.
3. Mature DCs were loaded with 25 uM peptide for 24 h at room temperature in Iscove's supplemented with 1% HS in the presence of 3 ug/ml human β_2-microglobulin (Sigma-Aldrich), washed twice with medium, and irradiated (40 Gy).

3.5.2.3. INDUCTION AND EXPANSION OF hTERT-SPECIFIC CTL CLONES

1. Peptide-loaded DCs ($1–2 \times 10^5$) were cultured for 10 days with 1×10^6 autologous CD8β-positive CTL precursors and 1×10^6 irradiated (80 Gy) autologous PBMC as feeders in 1.5 ml Yssel's medium *(35)* supplemented with 1% HS, 10 ng/ml IL-10 (R&D Systems, Oxon, UK), and 5 ng/ml IL-12 (R&D Systems) per well of a 24-well culture plate (Nunc, Intermed, Denmark).
2. On day 1, 10 ng/ml IL-10 (R&D Systems) was added to the culture. From day 10 on, CTL cultures were re-stimulated weekly with peptide-loaded mature DCs (1×10^5) in the presence of 5 ng/ml IL-7 (R&D Systems).

3. Two days after each re-stimulation, 20 U/ml IL-2 (Chiron, Amsterdam, The Netherlands) was added to the culture.
4. One day before each re-stimulation, a sample from each individual well was taken and analyzed by flow cytometry using the tetramers as indication.
5. Tetramer-positive CTLs were isolated by tetramer-directed flow sorting and cloned by limiting dilution, resulting routinely in >90% purity (*65*).
6. CTL clones were expanded in 24-well plates by weekly stimulation with an irradiated (80 Gy) feeder mix consisting of 1×10^6 allogenic PBMC from two different donors and 1×10^5 JY cells/ml of Yssel's medium supplemented with 1% HS, 100 ng/ml PHA (Murex Biotech, Dartfort, UK), and 20 U/ml IL-2 (*66*) (*see* **Note 3**).

3.6. Analysis of Cellular Cytotoxicity of the Induced hTERT-Specific CTLs

3.6.1. Chromium Release Assay

1. Cytolytic activity of the in vitro expanded CTL clones was analyzed in a standard 4-h ^{51}Cr-release assay to determine their ability to kill various target cells.
2. Target cells including tumor cells with various alleles, CD40-Bs, alone or peptide pulsed, were harvested from culture by a standard procedure, washed twice with PBS, and resuspended in RPMI-5 medium.
3. Target cells ($2 \times 10^{3-4}$) were labeled with 100 uCi $Na_2[^{51}Cr]O_4$ (Amersham, Bucks, UK) for 1 h at 37 °C and then washed extensively and added to the effector CTLs with various concentrations in the presence or absence of 1 uM antigenic peptide, in triplicate wells of a 96-well plate (Nunc).
4. After a 4-h incubation at 37 °C, radioactive content was measured in the supernatant to determine the percentage of specific lysis. Percent cytotoxicity is calculated as the [(cpm – spontaneous release)/(total cpm – spontaneous release)] $\times$ 100% in the chromium release assay.
5. Alternatively, JAM test can also be used to determine the cytotoxicity (*61,67*).

3.6.2. Intracellular IFN-γ Staining (ELISPOT)

1. The cytotoxicity of hTERT-specific CTLs can be tested on target cells for their ability to stimulate the release of TNF-γ from the target cells. Production of IFN-γ (R&D Systems) by the CTLs was determined using an ELISPOT assay as described previously.
2. 1×10^4 effector CTLs were incubated overnight with 1×10^4 target cells (loaded with or without 1 uM peptide) at 37 °C in a multiscreen 96-well filtration plate (Millipore) coated with mAB 1-D1K (Mabtech).
3. Plates were washed and incubated with biotinylated mAB 7-B6-1 (Mabtech) for 2–4 h at room temperature, followed by washing and incubation with streptavidin-alkaline phosphatase (Mabtech) for 1–2 h at room temperature.

4. IFN-γ spots were developed with an alkaline phosphatase conjugate substrate kit (Bio-Rad, Hercules, CA) and counted with an automated ELISPOT reader (Autoimmun Diagnostik, Strassberg, Germany).

3.6.3. Analysis of the Avidity of hTERT-Specific CTL Clones

Functional avidity involves the amount of TCR required for functional activity, thereby reflecting the amount of presented specific peptide/MHCs required for optimal CTL function. Analysis of functional avidity of hTERT-specific CTL clones was performed with target cells carrying with corresponding HLA alleles loaded with decreasing amounts of specific peptide in a chromium release assay. The half-maximal lysis by hTERT-specific CTL clones in the low nanomolar range reflects an intermediate avidity recognition of antigens *(66)*.

3.6.3.1. Cold Target Inhibition Assay *(32)*

1. To examine whether hTERT-specific CTLs lyse leukemia cells through the recognition of hTERT-peptides, which are actually processed in leukemia cells in the context of HLA-A24, cold target inhibition assays were performed.
2. Autologous B-LCLs established by the transformation of peripheral blood B lymphocytes using EBV were incubated with an hTERT-derived peptide at a concentration of 10 uM for 2 h.
3. After extensive washing, peptide-loaded cells were used as "cold" target cells. Various numbers of cold target cells were incubated with 1×10^5 cytotoxic effector cells for 1 h and then 1×10^4 ^{51}Cr-labeled leukemia cells were added to the wells.
4. Cytotoxicity was analyzed by chromium-release assay.

3.7. In Vitro Immunization

1. PBMCs were separated by centrifugation on Ficoll-Hypaque gradients and plated in 24-well plates at 5×10^5 cells/ml/well in RPMI 1640 medium supplemented with 10% human AB serum, L-glutamine, and antibiotics.
2. Autologous PBMCs (stimulators) were pulsed with hTERT synthetic peptides for 3 h at 37 °C.
3. Cells are then irradiated at 5000 rads, washed once, and added to the responder cells at a responder to stimulator ratio ranging between 1:1 and 1:4.
4. The next day, 12 units/ml IL-2 (Chiron) and 30 units/ml IL-7 (R&D systems) were added to the cultures.
5. Lymphocytes were re-stimulated weekly with peptide-pulsed autologous adherent cells as follows. First, autologous PBMCs were incubated with hTERT peptides (10 ug/ml) for 3 h at 37 °C. Non-adherent cells were removed by gentle wash and the adherent cells incubated with fresh medium containing the hTERT peptides

(10 ug/ml) for an additional 3 h at 37 °C. Second, responder cells from a previous stimulation cycle were harvested, washed, and added to the peptide-pulsed adherent cells at a concentration of 5×10^5 cells/ml (2 ml/well) in medium without peptide. Recombinant IL-2 and IL-7 were added to the cultures the next day.

3.8. In Vivo Immunization

3.8.1. Immunization of Murine

1. HHD transgenic mice, which express a chimeric HLA-A2.1/H2-D^b MHC class I molecule, are on a C57BL/6 background *(51)*. Mice were bred and maintained under specific pathogen-free conditions in the vivarium of the University of California at San Diego or the Institut Pasteur (Paris).
2. Humanized HLA-DR4 transgenic mice (HLA-DRB1*0401), which are murine class II deficient and transduced with human CD4 molecules, have been successfully used to identify human class II-restricted epitopes and to study immune responses *(52,53,68)*.
3. The HHD transgenic mice are immunized subcutaneously at the base of the tail with 100 ug individual hTERT peptides emulsified in incomplete Freund's adjuvant. Control group mice were injected with PBS emulsified in incomplete Freund's adjuvant.

3.8.2. Immunization of Human (Vaccination)

3.8.2.1. PATIENTS AND CLINICAL PROTOCOL *(69)*

1. Patients who had confirmed metastatic cancer were enrolled in the hTERT : I540 peptide vaccination protocol, which was approved by the Institutional Review Board of the National Cancer Institute.
2. Each patient underwent a complete clinical evaluation, including measurements and X-ray of all valuable tumor sites (*see* **Note 4**).
3. Patients were randomized into one of the following three cohorts of the study:
   a. hTERT : I540 peptide once a week every week for 4 or 10 cycles followed by a 3-week break and then repeated injections of peptide every week for 4 or 10 cycles (once weekly for 4 or 10 cycles);
   b. hTERT : I540 peptide once every 3 weeks; and
   c. hTERT : I540 peptide 4 days in a row every 3 weeks (Monday–Thursday). For each immunization, 1 mg peptide was injected as an emulsion with incomplete Freund's adjuvant (Montanide ISA-5, Seppic, Paris, France) in two equal volumes (1 ml/injection) into the subcutaneously tissue of the anterior thigh (*see* **Note 5**).

3.8.2.2. EVALUATION OF HLA-A*0201, hTERT, AND TELOMERASE EXPRESSION *(69)*

1. The expression of HLA-A2 was evaluated on all types of cells or tissues by fluorescence-activated cell sorting using an anti-HLA-A2 mAB (One lambda, Canoga Park, CA), and for some cells, DNA sequencing confirmed the presence of HLA-A*0201 (HLA Laboratory, National Institutes of Health).
2. The presence of hTERT mRNA in cells was assessed by reverse-transcriptase PCR using specific intron-spanning primers (forward GCCTGAGCTGTACTTTGTCAA and reverse CGCAAACAGCTTGTTCTCCATGTC).
3. Telomerase activity was measured by the telomeric repeat amplification protocol (TRAP) using TRAP*EZE* assay kit (Intergen, Gaithersburg, MD) in accordance with the manufacturer's instructions.

3.8.2.3. IN VITRO COMPARISON OF PEPTIDE AND TUMOR REACTIVITIES OF PRE-VACCINATION AND POST-VACCINATION IN PBMCS

1. PBMCs that had been cryopreserved before or after immunization with hTERT:I540 were stimulated with peptides *(70)*. T-cell cultures were established by plating PBMCs in 24-well plates (1.5×10^6 cells/ml, 2 ml/well) in culture medium containing 1–10 ug/ml peptide.
2. Two days later, 300 IU/ml recombinant IL-2 (Chiron) were added, and media were replaced as needed with fresh media containing IL-2.
3. Recognition of hTERT by bulk T-cell cultures was evaluated approximately 12 days after culture initiation on the basis of IFN-γ secretion in response to T2 cells pre-incubated with peptide and HLA-A*0201 hTERT cell lines.
4. T2 cells were incubated with peptide 1–3 h at 37°C and either used directly (peptide loaded) or washed twice before use (peptide pulsed).
5. 10^5 responder T cells were co-incubated with 10^5 stimulator cells (250 ul total) approximately 20 h at 37 °C, and the concentration of human IFN-γ in co-culture supernatants was measured using commercially available ELISA reagents (Endogen, Cambridge, MA).

3.9. Retroviral hTERT Transduction

1. The retroviral vector LZRS-hTERT-IRES-ΔNGFR was constructed by replacing the marker green fluorescent protein (GFP) for the coding sequence of a truncated signaling incompetent form of the low-affinity NGFR (ΔNGFR) *(71)* and inserting the hTERT coding sequence in the multiple cloning site in the original LZRS-MCS-IRES-GFP *(72)*.
2. The LZRS-hTERT-IRES-ΔNGFR construct was used to produce retroviral supernatant followed by retroviral hTERT transduction of CTL clones using protocols previously described *(70,72)*. 5×10^5 CTLs, stimulated for 48 h with feeder mix *(66)*, were resuspended in 0.5 ml retroviral supernatant supplemented with 20 U/ml

IL-2 and transferred to a fibronectin (RetroNectin, Takara, Otsu, Japan)-coated well of a non-tissue-culture-treated 24-well plate (BD Biosciences).

3. Plates were centrifuged for 90 min at 860 g, followed by 5-h incubation at 37 °C.
4. The CTLs were subsequently harvested, washed with culture medium, and kept at 37 °C overnight in culture medium supplemented with 20 U/ml IL-2.
5. The next day, retroviral transduction as described above was repeated.
6. The expression of hTERT was investigated after 48 h and later time points by flow-cytometric analysis of ΔNGFR marker gene expression using an NGFR-specific antibody.
7. CTLs transduced with hTERT were stimulated weekly or biweekly in 24-well plates with feeder mix.

4. Notes

1. All specimens were obtained following approval by Institutional Scientific Review Committee. Informed consent for blood donations was obtained from all volunteers. All donors including the healthy individuals and cancer patients were free of any kind of immunotherapy prior to the leukophresis.
2. The stability of each peptide bound to HLA-A2 is measured as the half-life of the complex. Dissociation of the test peptide from the HLA-A2.1 molecule reflects the half-life of fluorescence intensity of the peptide/MHC over time. The half-life of the complex refers to the time (h) required for a 50% reduction of the T_0 mean fluorescence intensity.
3. The preparation of CTL clones induced in vitro is often assessed for the phenotype and the purity by immunophenotypic analysis with immunofluorescence staining *(62,63)*. Dual-color FACS analysis using directly conjugated mAbs was performed to determine the surface expression.
4. All patients were confirmed to express HLA-A*0201 by high-resolution nested sequence PCR subtyping, and all patients signed an informed consent before treatment. No patient had received any treatment in the previous month, nor any of them received immunosuppressive drugs, including steroids. Before treatment, each patient underwent a leukapheresis, and PBMCs were cryopreserved at −180 °C after Ficoll-Hypaque separation.
5. Patients underwent leukapheresis 3 weeks after each course of immunization, and PBMCs were cryopreserved at −180 °C after Ficoll-Hypaque separation. For clinical administration, hTERT : I540 was purchased from Multiple Systems (San Diego, CA) as GMP-grade lyophilized powder.

Acknowledgment

The work was supported by grants from the Australia Research Council and National Health & Medical Research Council of Australia.

References

1. Van Pel, A., van der Bruggen, P., Coulie, P. G., Brichard, V. G., Lethe, B., van den Eynde, B., Uyttenhove, C., Renauld, J. C., and Boon, T. (1995) Genes coding for tumor antigens recognized by cytolytic T lymphocytes. *Immunol Rev* **145**, 229–50.
2. Rosenberg, S. A. (1997) Cancer vaccines based on the identification of genes encoding cancer regression antigens. *Immunol Today* **18**, 175–82.
3. Vonderheide, R. H., Hahn, W. C., Schultze, J. L., and Nadler, L. M. (1999) The telomerase catalytic subunit is a widely expressed tumor-associated antigen recognized by cytotoxic T lymphocytes. *Immunity* **10**, 673–9.
4. Rosenberg, S. A. (2001) Progress in human tumour immunology and immunotherapy. *Nature* **411**, 380–4.
5. Scardino, A., Gross, D. A., Alves, P., Schultze, J. L., Graff-Dubois, S., Faure, O., Tourdot, S., Chouaib, S., Nadler, L. M., Lemonnier, F. A., Vonderheide, R. H., Cardoso, A. A., and Kosmatopoulos, K. (2002) HER-2/neu and hTERT cryptic epitopes as novel targets for broad spectrum tumor immunotherapy. *J Immunol* **168**, 5900–6.
6. Lev, A., Denkberg, G., Cohen, C. J., Tzukerman, M., Skorecki, K. L., Chames, P., Hoogenboom, H. R., and Reiter, Y. (2002) Isolation and characterization of human recombinant antibodies endowed with the antigen-specific, major histocompatibility complex-restricted specificity of T cells directed toward the widely expressed tumor T-cell epitopes of the telomerase catalytic subunit. *Cancer Res* **62**, 3184–94.
7. Speiser, D. E., Cerottini, J. C., and Romero, P. (2002) Can hTERT peptide (540-548)-specific CD8 T cells recognize and kill tumor cells? *Cancer Immun* **2**, 14.
8. Maecker, B., von Bergwelt-Baildon, M. S., Anderson, K. S., Vonderheide, R. H., Anderson, K. C., Nadler, L. M., and Schultze, J. L. (2005) Rare naturally occurring immune responses to three epitopes from the widely expressed tumour antigens hTERT and CYP1B1 in multiple myeloma patients. *Clin Exp Immunol* **141**, 558–62.
9. Chen, L. (1998) Immunological ignorance of silent antigens as an explanation of tumor evasion. *Immunol Today* **19**, 27–30.
10. Ochsenbein, A. F., Klenerman, P., Karrer, U., Ludewig, B., Pericin, M., Hengartner, H., and Zinkernagel, R. M. (1999) Immune surveillance against a solid tumor fails because of immunological ignorance. *Proc Natl Acad Sci USA* **96**, 2233–8.
11. Lee, P. P., Yee, C., Savage, P. A., Fong, L., Brockstedt, D., Weber, J. S., Johnson, D., Swetter, S., Thompson, J., Greenberg, P. D., Roederer, M., and Davis, M. M. (1999) Characterization of circulating T cells specific for tumor-associated antigens in melanoma patients. *Nat Med* **5**, 677–85.
12. Ayyoub, M., Migliaccio, M., Guillaume, P., Lienard, D., Cerottini, J. C., Romero, P., Levy, F., Speiser, D. E., and Valmori, D. (2001) Lack of tumor

recognition by hTERT peptide 540-548-specific CD8(+) T cells from melanoma patients reveals inefficient antigen processing. *Eur J Immunol* **31**, 2642–51.

13. Van den Eynde, B. J., and van der Bruggen, P. (1997) T cell defined tumor antigens. *Curr Opin Immunol* **9**, 684–93.
14. Gilboa, E. (1999) The makings of a tumor rejection antigen. *Immunity* **11**, 263–70.
15. Schultze, J. L., and Vonderheide, R. H. (2001) From cancer genomics to cancer immunotherapy: toward second-generation tumor antigens. *Trends Immunol* **22**, 516–23.
16. Vonderheide, R. H. (2002) Telomerase as a universal tumor-associated antigen for cancer immunotherapy. *Oncogene* **21**, 674–9.
17. Minev, B., Hipp, J., Firat, H., Schmidt, J. D., Langlade-Demoyen, P., and Zanetti, M. (2000) Cytotoxic T cell immunity against telomerase reverse transcriptase in humans. *Proc Natl Acad Sci USA* **97**, 4796–801.
18. Vonderheide, R. H., Domchek, S. M., Schultze, J. L., George, D. J., Hoar, K. M., Chen, D. Y., Stephans, K. F., Masutomi, K., Loda, M., Xia, Z., Anderson, K. S., Hahn, W. C., and Nadler, L. M. (2004) Vaccination of cancer patients against telomerase induces functional antitumor CD8+ T lymphocytes. *Clin Cancer Res* **10**, 828–39.
19. Kosmatopoulos, K., Bolonaki, E., Cornet, S., Nikoloudi, E., Kanellou, P., Millaki, G., Miconnet, I., Christophillakis, C. H., Spiropoulou, M., Georgoulias, V., and Mavroudis, D. (2005) Safety and immunogenicity of the optimized cryptic peptide TERT572Y in patients with advanced malignancies: a phase I clinical study. *J Clin Oncol* **23**, 2005–579.
20. Kim, N. W., Piatyszek, M. A., Prowse, K. R., Harley, C. B., West, M. D., Ho, P. L., Coviello, G. M., Wright, W. E., Weinrich, S. L., and Shay, J. W. (1994) Specific association of human telomerase activity with immortal cells and cancer. *Science* **266**, 2011–5.
21. Shay, J. W., and Wright, W. E. (2002) Telomerase: a target for cancer therapeutics. *Cancer Cell* **2**, 257–65.
22. Hahn, W. C., Stewart, S. A., Brooks, M. W., York, S. G., Eaton, E., Kurachi, A., Beijersbergen, R. L., Knoll, J. H., Meyerson, M., and Weinberg, R. A. (1999) Inhibition of telomerase limits the growth of human cancer cells [see comments]. *Nat Med* **5**, 1164–70.
23. Greenberg, R. A., Chin, L., Femino, A., Lee, K. H., Gottlieb, G. J., Singer, R. H., Greider, C. W., and DePinho, R. A. (1999) Short dysfunctional telomeres impair tumorigenesis in the INK4a(delta2/3) cancer-prone mouse. *Cell* **97**, 515–25.
24. Herbert, B., Pitts, A. E., Baker, S. I., Hamilton, S. E., Wright, W. E., Shay, J. W., and Corey, D. R. (1999) Inhibition of human telomerase in immortal human cells leads to progressive telomere shortening and cell death. *Proc Natl Acad Sci USA* **96**, 14276–81.
25. Bayne, S., and Liu, J. P. (2005) Hormones and growth factors regulate telomerase activity in ageing and cancer. *Mol Cell Endocrinol* **240**, 11–22.

26. Parmiani, G., Castelli, C., Dalerba, P., Mortarini, R., Rivoltini, L., Marincola, F. M., and Anichini, A. (2002) Cancer immunotherapy with peptide-based vaccines: what have we achieved? Where are we going? *J Natl Cancer Inst* **94**, 805–18.

27. Nair, S. K., Heiser, A., Boczkowski, D., Majumdar, A., Naoe, M., Lebkowski, J. S., Vieweg, J., and Gilboa, E. (2000) Induction of cytotoxic T cell responses and tumor immunity against unrelated tumors using telomerase reverse transcriptase RNA transfected dendritic cells. *Nat Med* **6**, 1011–7.

28. Prowse, K. R., and Greider, C. W. (1995) Developmental and tissue-specific regulation of mouse telomerase and telomere length. *Proc Natl Acad Sci USA* **92**, 4818–22.

29. Hernandez, J., Garcia-Pons, F., Lone, Y. C., Firat, H., Schmidt, J. D., Langlade-Demoyen, P., and Zanetti, M. (2002) Identification of a human telomerase reverse transcriptase peptide of low affinity for HLA A2.1 that induces cytotoxic T lymphocytes and mediates lysis of tumor cells. *Proc Natl Acad Sci USA* **99**, 12275–80.

30. Vonderheide, R. H., Anderson, K. S., Hahn, W. C., Butler, M. O., Schultze, J. L., and Nadler, L. M. (2001) Characterization of HLA-A3-restricted cytotoxic T lymphocytes reactive against the widely expressed tumor antigen telomerase. *Clin Cancer Res* **7**, 3343–8.

31. Tokunaga, K., Ishikawa, Y., Ogawa, A., Wang, H., Mitsunaga, S., Moriyama, S., Lin, L., Bannai, M., Watanabe, Y., Kashiwase, K., Tanaka, H., Akaza, T., Tadokoro, K., and Juji, T. (1997) Sequence-based association analysis of HLA class I and II alleles in Japanese supports conservation of common haplotypes. *Immunogenetics* **46**, 199–205.

32. Arai, J., Yasukawa, M., Ohminami, H., Kakimoto, M., Hasegawa, A., and Fujita, S. (2001) Identification of human telomerase reverse transcriptase-derived peptides that induce HLA-A24-restricted antileukemia cytotoxic T lymphocytes. *Blood* **97**, 2903–7.

33. Vonderheide, R. H., Schultze, J. L., Anderson, K. S., Maecker, B., Butler, M. O., Xia, Z., Kuroda, M. J., von Bergwelt-Baildon, M. S., Bedor, M. M., Hoar, K. M., Schnipper, D. R., Brooks, M. W., Letvin, N. L., Stephans, K. F., Wucherpfennig, K. W., Hahn, W. C., and Nadler, L. M. (2001) Equivalent induction of telomerase-specific cytotoxic T lymphocytes from tumor-bearing patients and healthy individuals. *Cancer Res* **61**, 8366–70.

34. Harle-Bachor, C., and Boukamp, P. (1996) Telomerase activity in the regenerative basal layer of the epidermis inhuman skin and in immortal and carcinoma-derived skin keratinocytes. *Proc Natl Acad Sci USA* **93**, 6476–81.

35. Yasumoto, S., Kunimura, C., Kikuchi, K., Tahara, H., Ohji, H., Yamamoto, H., Ide, T., and Utakoji, T. (1996) Telomerase activity in normal human epithelial cells. *Oncogene* **13**, 433–9.

36. Kolquist, K. A., Ellisen, L. W., Counter, C. M., Meyerson, M., Tan, L. K., Weinberg, R. A., Haber, D. A., and Gerald, W. L. (1998) Expression of TERT in

early premalignant lesions and a subset of cells in normal tissues [see comments]. *Nat Genet* **19**, 182–6.

37. Norrback, K. F., and Roos, G. (1997) Telomeres and telomerase in normal and malignant haematopoietic cells. *Eur J Cancer* **33**, 774–80.

38. Masutomi, K., Yu, E. Y., Khurts, S., Ben-Porath, I., Currier, J. L., Metz, G. B., Brooks, M. W., Kaneko, S., Murakami, S., DeCaprio, J. A., Weinberg, R. A., Stewart, S. A., and Hahn, W. C. (2003) Telomerase maintains telomere structure in normal human cells. *Cell* **114**, 241–53.

39. Dhodapkar, M. V., Steinman, R. M., Sapp, M., Desai, H., Fossella, C., Krasovsky, J., Donahoe, S. M., Dunbar, P. R., Cerundolo, V., Nixon, D. F., and Bhardwaj, N. (1999) Rapid generation of broad T-cell immunity in humans after a single injection of mature dendritic cells. *J Clin Invest* **104**, 173–80.

40. Banchereau, J., Palucka, A. K., Dhodapkar, M., Burkeholder, S., Taquet, N., Rolland, A., Taquet, S., Coquery, S., Wittkowski, K. M., Bhardwaj, N., Pineiro, L., Steinman, R., and Fay, J. (2001) Immune and clinical responses in patients with metastatic melanoma to CD34(+) progenitor-derived dendritic cell vaccine. *Cancer Res* **61**, 6451–8.

41. Gilliet, M., Kleinhans, M., Lantelme, E., Schadendorf, D., Burg, G., and Nestle, F. O. (2003) Intranodal injection of semimature monocyte-derived dendritic cells induces T helper type 1 responses to protein neoantigen. *Blood* **102**, 36–42.

42. Jonuleit, H., Giesecke-Tuettenberg, A., Tuting, T., Thurner-Schuler, B., Stuge, T. B., Paragnik, L., Kandemir, A., Lee, P. P., Schuler, G., Knop, J., and Enk, A. H. (2001) A comparison of two types of dendritic cell as adjuvants for the induction of melanoma-specific T-cell responses in humans following intranodal injection. *Int J Cancer* **93**, 243–51.

43. De Vries, I. J., Krooshoop, D. J., Scharenborg, N. M., Lesterhuis, W. J., Diepstra, J. H., Van Muijen, G. N., Strijk, S. P., Ruers, T. J., Boerman, O. C., Oyen, W. J., Adema, G. J., Punt, C. J., and Figdor, C. G. (2003) Effective migration of antigen-pulsed dendritic cells to lymph nodes in melanoma patients is determined by their maturation state. *Cancer Res* **63**, 12–7.

44. Nestle, F. O., Alijagic, S., Gilliet, M., Sun, Y., Grabbe, S., Dummer, R., Burg, G., and Schadendorf, D. (1998) Vaccination of melanoma patients with peptide- or tumor lysate-pulsed dendritic cells. *Nat Med* **4**, 328–32.

45. Dhodapkar, M. V., Steinman, R. M., Krasovsky, J., Munz, C., and Bhardwaj, N. (2001) Antigen-specific inhibition of effector T cell function in humans after injection of immature dendritic cells. *J Exp Med* **193**, 233–8.

46. Dudley, M. E., Wunderlich, J. R., Robbins, P. F., Yang, J. C., Hwu, P., Schwartzentruber, D. J., Topalian, S. L., Sherry, R., Restifo, N. P., Hubicki, A. M., Robinson, M. R., Raffeld, M., Duray, P., Seipp, C. A., Rogers-Freezer, L., Morton, K. E., Mavroukakis, S. A., White, D. E., and Rosenberg, S. A. (2002) Cancer regression and autoimmunity in patients after clonal repopulation with antitumor lymphocytes. *Science* **298**, 850–4.

47. DiBrino, M., Parker, K. C., Shiloach, J., Knierman, M., Lukszo, J., Turner, R. V., Biddison, W. E., and Coligan, J. E. (1993) Endogenous peptides bound to HLA-A3 possess a specific combination of anchor residues that permit identification of potential antigenic peptides. *Proc Natl Acad Sci USA* **90**, 1508–12.

48. DiBrino, M., Tsuchida, T., Turner, R. V., Parker, K. C., Coligan, J. E., and Biddison, W. E. (1993) HLA-A1 and HLA-A3 T cell epitopes derived from influenza virus proteins predicted from peptide binding motifs. *J Immunol* **151**, 5930–5.

49. Tsomides, T. J., Walker, B. D., and Eisen, H. N. (1991) An optimal viral peptide recognized by CD8+ T cells binds very tightly to the restricting class I major histocompatibility complex protein on intact cells but not to the purified class I protein. *Proc Natl Acad Sci USA* **88**, 11276–80.

50. Tanaka, F., Fujie, T., Tahara, K., Mori, M., Takesako, K., Sette, A., Celis, E., and Akiyoshi, T. (1997) Induction of antitumor cytotoxic T lymphocytes with a MAGE-3-encoded synthetic peptide presented by human leukocytes antigen-A24. *Cancer Res* **57**, 4465–8.

51. Pascolo, S., Bervas, N., Ure, J. M., Smith, A. G., Lemonnier, F. A., and Perarnau, B. (1997) HLA-A2.1-restricted education and cytolytic activity of CD8(+) T lymphocytes from beta2 microglobulin (beta2m) HLA-A2.1 monochain transgenic H-2Db beta2m double knockout mice. *J Exp Med* **185**, 2043–51.

52. Congia, M., Patel, S., Cope, A. P., De Virgiliis, S., and Sonderstrup, G. (1998) T cell epitopes of insulin defined in HLA-DR4 transgenic mice are derived from preproinsulin and proinsulin. *Proc Natl Acad Sci USA* **95**, 3833–8.

53. Sonderstrup, G., and McDevitt, H. (1998) Identification of autoantigen epitopes in MHC class II transgenic mice. *Immunol Rev* **164**, 129–38.

54. Sonderstrup, G., Cope, A. P., Patel, S., Congia, M., Hain, N., Hall, F. C., Parry, S. L., Fugger, L. H., Michie, S., and McDevitt, H. O. (1999) HLA class II transgenic mice: models of the human CD4+ T-cell immune response. *Immunol Rev* **172**, 335–43.

55. Ruppert, J., Sidney, J., Celis, E., Kubo, R. T., Grey, H. M., and Sette, A. (1993) Prominent role of secondary anchor residues in peptide binding to HLA-A2.1 molecules. *Cell* **74**, 929–37.

56. Krausa, P., Brywka, M., 3rd, Savage, D., Hui, K. M., Bunce, M., Ngai, J. L., Teo, D. L., Ong, Y. W., Barouch, D., Allsop, C. E., and et al. (1995) Genetic polymorphism within HLA-A*02: significant allelic variation revealed in different populations. *Tissue Antigens* **45**, 223–31.

57. Parker, K. C., Bednarek, M. A., and Coligan, J. E. (1994) Scheme for ranking potential HLA-A2 binding peptides based on independent binding of individual peptide side-chains. *J Immunol* **152**, 163–75.

58. Firat, H., Garcia-Pons, F., Tourdot, S., Pascolo, S., Scardino, A., Garcia, Z., Michel, M. L., Jack, R. W., Jung, G., Kosmatopoulos, K., Mateo, L., Suhrbier, A., Lemonnier, F. A., and Langlade-Demoyen, P. (1999) H-2 class I knockout,

HLA-A2.1-transgenic mice: a versatile animal model for preclinical evaluation of antitumor immunotherapeutic strategies. *Eur J Immunol* **29**, 3112–21.

59. Altman, J. D., Moss, P. A., Goulder, P. J., Barouch, D. H., McHeyzer-Williams, M. G., Bell, J. I., McMichael, A. J., and Davis, M. M. (1996) Phenotypic analysis of antigen-specific T lymphocytes. *Science* **274**, 94–6.

60. Kannagi, M., Shida, H., Igarashi, H., Kuruma, K., Murai, H., Aono, Y., Maruyama, I., Osame, M., Hattori, T., Inoko, H., and et al. (1992) Target epitope in the Tax protein of human T-cell leukemia virus type I recognized by class I major histocompatibility complex-restricted cytotoxic T cells. *J Virol* **66**, 2928–33.

61. Schultze, J. L., Seamon, M. J., Michalak, S., Gribben, J. G., and Nadler, L. M. (1997) Autologous tumor infiltrating T cells cytotoxic for follicular lymphoma cells can be expanded in vitro. *Blood* **89**, 3806–16.

62. Schultze, J. L., Cardoso, A. A., Freeman, G. J., Seamon, M. J., Daley, J., Pinkus, G. S., Gribben, J. G., and Nadler, L. M. (1995) Follicular lymphomas can be induced to present alloantigen efficiently: a conceptual model to improve their tumor immunogenicity. *Proc Natl Acad Sci USA* **92**, 8200–4.

63. Schultze, J. L., Michalak, S., Seamon, M. J., Dranoff, G., Jung, K., Daley, J., Delgado, J. C., Gribben, J. G., and Nadler, L. M. (1997) CD40-activated human B cells: an alternative source of highly efficient antigen presenting cells to generate autologous antigen-specific T cells for adoptive immunotherapy. *J Clin Invest* **100**, 2757–65.

64. Romani, N., Reider, D., Heuer, M., Ebner, S., Kampgen, E., Eibl, B., Niederwieser, D., and Schuler, G. (1996) Generation of mature dendritic cells from human blood. An improved method with special regard to clinical applicability. *J Immunol Methods* **196**, 137–51.

65. Yssel, H., De Vries, J. E., Koken, M., Van Blitterswijk, W., and Spits, H. (1984) Serum-free medium for generation and propagation of functional human cytotoxic and helper T cell clones. *J Immunol Methods* **72**, 219–27.

66. Schreurs, M. W., Scholten, K. B., Kueter, E. W., Ruizendaal, J. J., Meijer, C. J., and Hooijberg, E. (2003) In vitro generation and life span extension of human papillomavirus type 16-specific, healthy donor-derived CTL clones. *J Immunol* **171**, 2912–21.

67. Matzinger, P. (1991) The JAM test. A simple assay for DNA fragmentation and cell death. *J Immunol Methods* **145**, 185–92.

68. Geluk, A., Taneja, V., van Meijgaarden, K. E., de Vries, R. R., David, C. S., and Ottenhoff, T. H. (1998) HLA-DR/DQ transgenic, class II deficient mice as a novel model to select for HSP T cell epitopes with immunotherapeutic or preventative vaccine potential. *Biotherapy* **10**, 191–6.

69. Parkhurst, M. R., Riley, J. P., Igarashi, T., Li, Y., Robbins, P. F., and Rosenberg, S. A. (2004) Immunization of patients with the hTERT:540-548 peptide induces peptide-reactive T lymphocytes that do not recognize tumors endogenously expressing telomerase. *Clin Cancer Res* **10**, 4688–98.

70. Rosenberg, S. A., Yang, J. C., Schwartzentruber, D. J., Hwu, P., Marincola, F. M., Topalian, S. L., Restifo, N. P., Dudley, M. E., Schwarz, S. L., Spiess, P. J., Wunderlich, J. R., Parkhurst, M. R., Kawakami, Y., Seipp, C. A., Einhorn, J. H., and White, D. E. (1998) Immunologic and therapeutic evaluation of a synthetic peptide vaccine for the treatment of patients with metastatic melanoma. *Nat Med* **4**, 321–7.

71. Bordignon, C., Notarangelo, L. D., Nobili, N., Ferrari, G., Casorati, G., Panina, P., Mazzolari, E., Maggioni, D., Rossi, C., Servida, P., Ugazio, A. G., and Mavilio, F. (1995) Gene therapy in peripheral blood lymphocytes and bone marrow for ADA-immunodeficient patients. *Science* **270**, 470–5.

72. Heemskerk, M. H., Hooijberg, E., Ruizendaal, J. J., van der Weide, M. M., Kueter, E., Bakker, A. Q., Schumacher, T. N., and Spits, H. (1999) Enrichment of an antigen-specific T cell response by retrovirally transduced human dendritic cells. *Cell Immunol* **195**, 10–7.

8

Establishing Cell-Based Reporter Systems for the Analysis of hTERT Expression

Yi-Yuan Huang, Jing-Wen Shih, and Jing-Jer Lin

Summary

Telomeres are the protective structures at the end of eukaryotic chromosomes. Telomerase is a ribonucleoprotein that contains both an RNA and a protein component for the maintenance of telomere length. Telomerase activity is detected in the majority of malignant tumors, but not in normal somatic cells, suggesting that telomerase reactivation is a crucial step in cell immortality and carcinogenesis. The mechanism of how telomerase is activated during tumorigenesis remains unclear. However, the expression of the human telomerase reverse transcriptase (*hTERT*) gene, which encodes the catalytic protein subunit of human telomerase, has been shown to be the major determining factor. To gain insight into the mechanisms regulating hTERT expression and to facilitate the screening of agents that affect hTERT expression, we have established cell-based systems for monitoring hTERT expression. We linked the hTERT promoter to two different reporter genes encoding green fluorescence protein (GFP) and secreted alkaline phosphatase (SEAP), respectively. These constructs were then transfected into H1299 and hTERT-BJ1 cells. Stable clones harboring these DNA constructs were isolated. In these cells, hTERT expression can be monitored through the quantification of GFP or SEAP activity on an automatic plate reader. Using these systems, we have identified several small molecule compounds that affect the expression of telomerase.

Key Words: Telomerase; telomere; hTERT; SEAP; GFP.

1. Introduction

Telomeres, the ends of eukaryotic chromosomes, are essential for the stability and replication *(1,2)*. In human chromosomes, telomeres consist of

From: *Methods in Molecular Biology, vol. 405: Telomerase Inhibition*
Edited by: L. G. Andrews and T. O. Tollefsbol © Humana Press Inc., Totowa, NJ

approximately 15 kbp T_2AG_3 tandem repeats *(3)*. Telomeres are essential for maintaining chromosome integrity. They protect chromosomes from degradation by nucleases, facilitate complete replication of chromosomes, and differentiate linear chromosome ends from broken ends. During each cell division, telomeres are progressively shortened as a result of the incomplete lagging strand replication. A reduction in the telomere length to a critical level can lead to an irreversible growth arrest called cellular senescence. In contrast, the telomere length in cancer cells is maintained by telomerase and leads to cellular immortalization *(4,5)*. Because of the important role of telomerase in cellular immortalization, telomerase has been the focus for anti-cancer drug developments *(6)*. Agents that target telomerase activity using antisense oligonucleotides or small molecule compounds have been proven to be effective in inhibiting telomerase activity *(7)*.

Several genes have been shown to possess different expression profiles in tumors and normal cells. For example, the tyrosine gene promoter in melanomas *(8)*, carcinoembryonic antigen (CEA) promoter in colorectal and lung cancer cells *(9)*, the mucin1 (MUC1) promoter in breast cancer *(10)*, and the E2F promoter in cancers with defective retinoblastoma gene *(11)* are more active in these cancer cells. Promoters of these genes have been evaluated for their potential in cancer treatments. Telomerase is a ribonucleoprotein that is composed of a template-containing RNA subunit, human telomerase RNA component (hTR), and catalytic protein subunit human telomerase reverse transcriptase (hTERT) in humans. The expression of hTERT appears to be the key determinant of telomerase activity. It was shown that the regulation of hTERT expression could occur at different levels *(12)*. Nevertheless, the transcriptional control has been the major factor in regulating hTERT expression. Using a reporter system, it was shown that the hTERT promoter is activated in cancer cell lines but is repressed in normal primary cells *(13,14)*. Thus, even though the mechanism of how hTERT is inactivated during development and reactivated during tumorigenesis is still unclear, an additional approach to target telomerase for anti-cancer drugs development is to affect hTERT expression.

Here, we discuss the use of hTERT promoter in developing the cell-based systems to monitor telomerase expression. We have established both normal and cancer cell lines harboring reporter genes, green fluorescence protein (GFP), and secreted alkaline phosphatase (SEAP), under the control of hTERT promoter. Using these systems, we have identified several small molecule compounds that affect the expression of telomerase *(14–16)*.

2. Materials

2.1. Plasmid Constructions

1. pP$_{CMV}$-GFP: Plasmid pP$_{CMV}$-GFP was constructed by inserting a 732-bp *Not*I-digested GFP fragment from pGreenLantern (Gibco BRL, Gaithersburg, MD, USA) downstream to the cytomegalovirus (CMV) promoter of pBK-CMV (Stratagene, La Jolla, CA, USA).

2. pP$_{CMV}$-SEAP: Plasmid pP$_{CMV}$-SEAP was constructed by inserting a 1.7-kbp *Eco*RI–*Kpn*I SEAP fragment, which was polymerase chain reaction (PCR) amplified from pSEAP2-control (BD Biosciences Clontech, Palo Alto, CA, USA), into the *Eco*RI-digested and *Kpn*I-digested pCMV-GFP. Primers used to amplify SEAP fragments were SEAP-F (5′-CGCGAATTCGCCCACCATGCTGC-3′) and SEAP-R (5′-CTGGTACCATGCAATTGTTGTTAAC-3′).

3. pP$_{TERT}$-GFP: There were two steps in constructing pP$_{TERT}$-GFP plasmid. The first step was to cut off CMV promoter from pCMV-GFP to generate pGFP. It was conducted by digesting pCMV-GFP with *Pst*I and *Nsi*I, blunted with T4 DNA polymerase, and self-ligated. The second step was to insert a 3.4-kbp hTERT promoter fragment into the *Sal*I-digested and *Spe*I-digested pGFP. The hTERT promoter was PCR amplified from H1299 genomic DNA using forward primer (5′-GGGTCGACTACCTGCAGGCCCGAAAAG-3′) and reverse primer (5′-GGACTAGTCTTCCCACGTGCGCAGCAGGA-3′).

4. pP$_{TERT}$-SEAP: Plasmid pP$_{TERT}$-SEAP was constructed by replacing the GFP of pP$_{TERT}$-GFP with a 1.7-kbp PCR-amplified SEAP fragment using method similar to that described for pP$_{CMV}$-SEAP construction.

2.2. Cell Culture

1. Non-small cell lung cancer H1299 cells: RPMI 1640 medium (Gibco BRL) supplemented with 10% fetal bovine serum, 100 units/ml penicillin, and 100 mg/ml streptomycin in a humidified atmosphere with 5% CO_2 at 37 °C.

2. The hTERT-immortalized human skin fibroblast hTERT-BJ1 (BD Biosciences Clontech, Palo Alto, CA, USA),: Dulbecco's modified Eagle's medium (DMEM) (Gibco BRL) supplemented with 10% fetal calf serum, 100 units/ml penicillin, 100 mg/ml streptomycin, 1 mM sodium pyruvate (Gibco BRL), and 4 mM L-arginine (final concentration) in humidified atmosphere with 5% CO_2 at 37 °C.

3. Trypsin (0.25%, w/v) for 100 ml: 0.25 g trypsin (Sigma, St. Louis, MO, USA), 0.02 g ethylenediamine tetraacetic acid (EDTA) (Merck, Darmstadt, Germany), and 0.05 g glucose (Sigma) in 100 ml 1× phosphate-buffered saline (PBS). Adjust to pH 7.0 with HCl. Sterilize by passing the solution through a 0.22-μm filter (Pall Life Sciences, Ann Arbor, MI, USA), stored at −20°C.

4. PBS buffer (10×): 80 g NaCl, 2 g KCl, 6 g Na_2HPO_4, and 2 g KH_2PO_4 in 1 l distilled water, adjust to pH 7.4 with HCl. Autoclave and store at room temperature.

Dilute one part of 10× PBS buffer with nine parts of water to make the working solution.

2.3. MTT Reagent and Lysis Buffer

1. Dissolve 3-(4,5-di-methylthiazol)-2,5-diphenyltetrazolium bromide (MTT) (USB/ Amersham Life Science, Cleveland, OH, USA) at 5 mg/ml in PBS, pass through 0.22-μm filter (Pall Life Sciences), and store at 4 °C.
2. MTT lysis buffer for cell lysis: 20% (w/v) sodium dodecyl sulfate (SDS) in 50% (w/v) N,N-dimethylformamide (DMF). Store at room temperature and pre-warm to 37 °C before using.

2.4. Reagents for SEAP Assays

1. SEAP buffer (2×): 2 M diethanolamine, 1 mM $MgCl_2$, and 20 mM L-homoarginine in distilled water. Store at 4 °C and pre-warm to 37 °C before using.
2. p-nitrophenyl phosphate buffer: 120 mM p-nitrophenyl phosphate (Sigma) in distilled water.

2.5. Fluorescence of GFP Expression

1. Microscope cover-slips (12 mm, Assistant, Hecht, Sondheim, Germany).
2. Mounting solution: 20 mM n-propylgallate in 90% glycerol (one part 1× PBS added to nine parts 100% glycerol). Sonication is required to dissolve the n-propylgallate. Store at 4 °C and pre-warm to 37 °C before using.
3. Paraformaldehyde: 4% (w/v) paraformaldehyde in PBS. To dissolve paraformaldehyde requires temperature at 65 °C. Cool to room temperature, store at −20 °C, and pre-warm to 37 °C before using.

3. Methods

Because the expression of human telomerase catalytic component is the key regulator in telomerase activity, we analyzed the expression of telomerase by monitoring the expression of hTERT as the criteria. An approximately 3.4-kbp DNA fragment ranging from −3338 to +1 bp of the *hTERT* gene was PCR amplified from genomic DNA and subcloned upstream to a *GFP* or an *SEAP* gene. The resulting plasmids were transfected into H1299 or hTERT-BJ1 by electroporation (*see* **Fig. 1**). Cell lines with the expression of hTERT monitored by either GFP or SEAP reporter systems were established by selection using G418 at a concentration of 0.8 mg/ml. These H1299-derived or hTERT-BJ1-derived stable clones were cultured using conditions that are similar to their parental cells. We also generated the CMV promoter-driven GFP or SEAP reporter constructs as a positive control and GFP without promoter as a negative

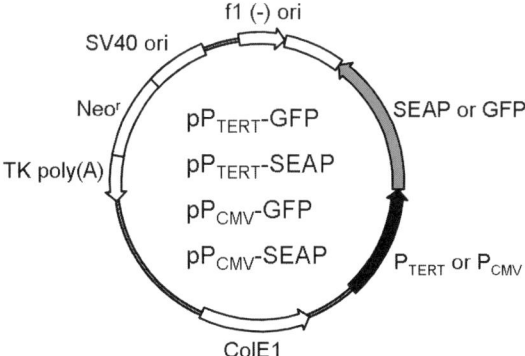

Fig. 1. Plasmid constructs used to generate cell-based human telomerase reverse transcriptase (hTERT) reporting systems.

control. The behavior of hTERT promoter in H1299 or hTERT-BJ1 cells is consistent with its expression in cancer cells and low expressions in normal cells *(14)*. For example, the *GFP* reporter gene expresses under the control of hTERT or CMV promoter give positive signals, whereas promoter-less GFP gives only background signals under fluorescence microscope (*see* **Fig. 2** and **Note 1**). Using this system, small molecule compounds that activate hTERT in normal cells and repress hTERT in cancer cells were identified *(14–16)*. Taking **Fig. 3** as an example, we have identified compound **Bb** that selectively represses the expression of hTERT promoter-driven SEAP and GFP without affecting the expression of CMV promoter-driven genes. Consistent with results from the reporter systems, the endogenous hTERT mRNA level was decreased upon compound **Bb** treatments (data not shown). Thus, our systems provide a useful tool to monitor the expression of hTERT and to screen for agents that affect hTERT expressions.

3.1. Preparation of Samples for GFP System *(See Note 2)*

1. Grow cells harboring pP_{TERT}-GFP to near confluence and harvest by treating with 0.25% trypsin. Count the cells and then seed approximately 10^6 cells to 100-mm culture dishes. Incubate the cells at 37 °C for 24 h.
2. Change the cells with fresh media before compound treatments.
3. Add varying amounts of tested compounds with a volume less than one tenth of the medium and then incubate for another 24 h.
4. Wash the treated cells twice with PBS. Add 1.5 ml 0.25% trypsin, enough to cover the cells, and incubate at 37 °C for 5–10 min to detach the cells from culture plates.

Light   Fluorescence
microscopy   microscopy

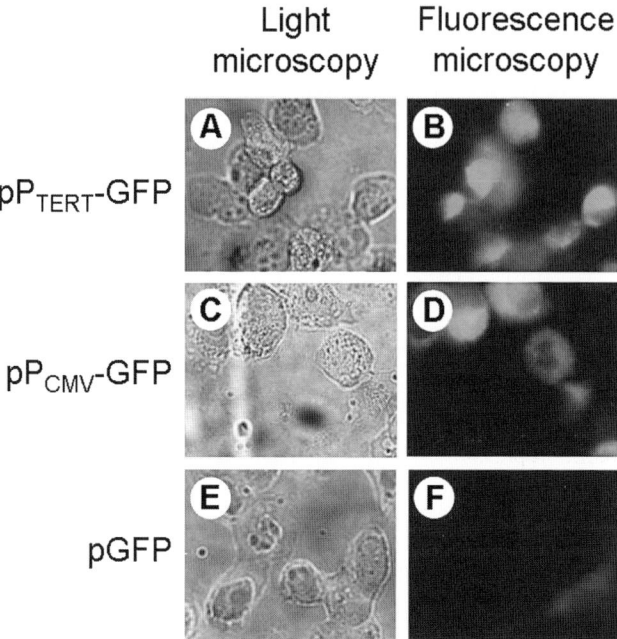

Fig. 2. Observation of green fluorescence protein (GFP) expression in pP_{TERT}-GFP reporter clones. H1299 cells, a telomerase positive cell line, were transfected with pP_{TERT}-GFP (**A** and **B**), pP_{CMV}-GFP (**C** and **D**), or pGFP (**E** and **F**), respectively. Stable clones were selected. Left and right panels were images taken by light and fluorescence microscopy, respectively.

5. Add 10 ml PBS to the cells and transfer them to 15-ml centrifuge tubes. Harvest the cells by centrifuging at 300 g (KN-70, Kubota, Tokyo, Japan) for 10 min. Remove the supernatants and re-suspend the cells in PBS (*see* **Note 3**).
6. Seed 5×10^5 cells into 96-well Costar opaque plates and quantify the GFP fluorescence using Labsystem Fluoroskan Ascent and the software Appfwgl (*see* **Notes 4–6**).

3.2. Preparation of Samples for SEAP Assay

1. Grow cells harboring pP_{TERT}-SEAP to near confluence and harvest the cells following the protocol described in **Subheading 3.1**. Transfer 2×10^3 cells in 180-µl culture medium to 96-well plates and incubate at 37 °C for 24 h (*see* **Note 7**).
2. Change the cells with fresh media before compound treatments.
3. Add 20 µl test compounds to the cells and incubate at 37 °C for another 24 h.

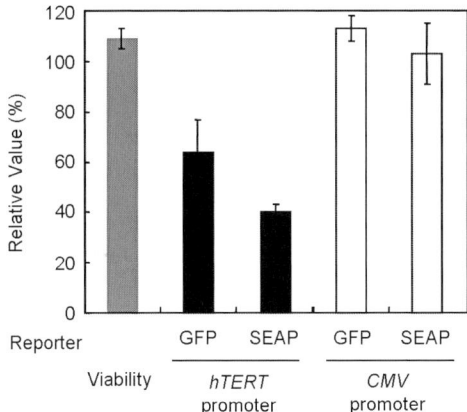

Fig. 3. Specific repression of human telomerase reverse transcriptase (hTERT) promoter by **Bb**. H1299 cells harboring hTERT promoter-driven secreted alkaline phosphatase (SEAP) or green fluorescence protein (GFP) or cytomegalovirus (CMV) promoter-driven SEAP or GFP were treated with **Bb** for 24 h. The 3-(4,5-di-methylthiazol)-2,5-diphenyltetrazolium bromide (MTT) tests, SEAP activity, and the GFP intensity were then analyzed according to the procedures described in the text.

4. Transfer 150 μl culture medium containing SEAP to a new 96-well plate and incubate at 65 °C for 10 min to deactivate heat-labile phosphatase.
5. Wash the remaining cells with 100 μl PBS twice for MTT analysis (described in **Subheading 3.3.**).
6. Transfer 50 μl medium to another new 96-well plate.
7. Add an equal amount of 2× SEAP buffer (50 μl) to the media followed by 10 μl 120 mM *p*-nitrophenyl phosphate. Mix them completely and take the absorptions at 405 nm using ELISA reader (BIO-RAD Model 550 Microplate reader, Bio-Rad, Hercules, CA, USA). Determine the SEAP activities as the rates of absorption change (*see* **Note 8**).

3.3. MTT Assays

1. The tetrazolium reagent (MTT, USB/Amersham Life Science) was designed to yield a colored formazan upon metabolic reduction by viable cells (*17*).
2. Seed approximately 2×10^3 cells in 180-μl culture media onto 96-well plates and incubate at 37 °C for 24 h.
3. Add 20 μl test compounds to the culture medium and incubate the cells for another 24 h.

4. After 24 h, remove the medium and wash the cells with 100 μl PBS. Add 100 μl serum-free medium and 25 μl 5 mg/ml MTT reagent to cells and incubate at 37 °C for 4 h.
5. Add 100 μl MTT lysis buffer to cells and incubate at 37 °C for another 16 h.
6. Measure the absorbency at 550 nm using an ELISA reader (BIO-RAD Model 550 Microplate reader).

3.4. Fluorescent Microscope Analysis for GFP Expression Cells

1. Preparation of cover-slips: soak cover-slips (Assistant Germany) in 0.1 N HCl overnight, wash with running water for 3 h, and distilled water 3×. Autoclave them.
2. Place the cover-slips on a 24-well plate, added 250–300 μl poly-D-lysine (PDL) to cover these cover-slips, and at 4 °C overnight.
3. Rinse these cover-slips with serum-free medium.
4. Seed approximately 10^4 cells onto the cover-slips. Place them in the incubator at 37 °C for 20 min. Add more medium when the cells are attached to the cover-slips.
5. Gently wash the cells with PBS twice. Add 4% paraformaldehyde to fix the cells on the cover-slips for 30 min.
6. Place cover-slips upside down on the glass plates with mounting solution and then seal them with nail oil.
7. Take the images using fluorescence microscope (PM-30, Olympus, Tokyo, Japan).

4. Notes

1. In general, the CMV promoter-driven reporter system gave a higher expression level than that of hTERT promoter in H1299 cells. To have a better comparison between these two promoters, we have picked clones that gave similar expression levels of the reporter genes for our experiments. In some experiments, the results were also confirmed using different stable clones.
2. Because the GFP fluorescence measurements were affected by a variety of factors including plates, temperatures, pH, and culture medium, quantification of GFP systems is not always satisfying. We found the SEAP system is easier and cheaper to perform. It is suitable for large-scale screening for agents that affect hTERT expression.
3. Owing to the high-fluorescence background of regular culture medium, we usually wash the cells with PBS and then take the GFP readings in PBS.
4. We have used Labsystem Fluoroskan Ascent for the quantification of GFP fluorescence. The opaque plates were used in our analysis. Even though most suppliers claim their opaque plates could be used for GFP quantifications, we found only two brands gave satisfying readings. In our system, we typically use Costar 3912 white opaque plate for the experiments. The NUNC 136101 white opaque plate also works for us. However, because these plates are not designed for cell culture, we usually culture the cells in regular culture plates and then transfer the cells to opaque plates before taking the GFP readings.

5. In GFP system, a lower temperature (i.e., 4 °C) gives higher readings than that at room temperature. However, because our fluorometer is not equipped with a temperature control unit, samples in low temperature gradually increase temperature leading to the final reading displaying high deviations. Thus, we usually take the GFP reading directly at room temperature.

6. Even though there is clonal variation of GFP expression in different stable clones, we have found the liner range for GFP reading is within 4×10^4 to 1×10^6 cells/well.

7. The liner range for a typical assay is from 2×10^3 to 1×10^4 cells, suitable for most 96-well plates for large-scale analysis. However, because SAEP is secreted into the culture medium, the effect of testing agents on phosphatase activity has to be checked. We routinely use the SEAP system as the first-line screening system and the GFP system as the second-line confirm system.

8. For a better absorption reading using ELISA reader, it is necessary to clean the bubbles on the top of the media to be sure of not affecting the results.

Acknowledgments

The authors thank Yen-Ru Pan, Chen-Yi Wang, and Yu-Cheng Lu for development of the cell-based systems. This work is supported by National Science council grant NSC94-3112-B-010-020 and National Health Research Institute grant NHRI-EX94-9436SI.

References

1. Blackburn, E. H. (2005) Telomeres and telomerase: their mechanisms of action and the effects of altering their functions. *FEBS Lett.* **579**, 859–862.

2. Vega, L. R., Mateyak, M. K., Zakian, V. A. (2003) Getting to the end: telomerase access in yeast and humans. *Nat. Rev. Mol. Cell. Biol.* **4**, 948–959.

3. Moyzis, R. K., Buckingham, J. M., Cram, L. S., Dani, M., Deaven, L. L., Jones, M. D., Meyne, J., Ratliff, R. L., Wu, J. R. (1988) A highly conserved repetitive DNA sequence, (TTAGGG)n, present at the telomeres of human chromosomes. *Proc. Natl. Acad. Sci. U. S. A.* **85**, 6622–6626.

4. Kim, N. W., Piatyszek, M. A., Prowse, K. R., Harley, C. B., West, M. D., Ho, P. L., Coviello, G. M., Wright, W. E., Weinrich, S. L., Shay, J. W. (1994) Specific association of human telomerase activity with immortal cells and cancer. *Science* **266**, 2011–2015.

5. Hahn, W. C., Counter, C. M., Lundberg, A. S., Beijersbergen, R. L., Brooks, M. W., Weinberg, R. A. (1999) Creation of human tumour cells with defined genetic elements. *Nature* **400**, 464–468.

6. Shay, J. W., Wright, W. E. (2002) Telomerase: a target for cancer therapeutics. *Cancer Cell* **2**, 257–265.

7. Hsu, Y.-H., Lin, J.-J. (2005) Telomere and telomerase as targets for anti-cancer and regeneration therapies. *Acta Pharmacol. Sin.* **26**, 513–518.

8. Vile, R. G., Hart, I. R. (1993) In vitro and in vivo targeting of gene expression to melanoma cells. *Cancer Res.* **53**, 962–967.
9. Osaki, T., Tanio, Y., Tachibana, I., Hosoe, S., Kumagai, T., Kawase, I., Oikawa, S., Kishimoto, T. (1994) Gene therapy for carcinoembryonic antigen-producing human lung cancer cells by cell type-specific expression of herpes simplex virus thymidine kinase gene. *Cancer Res.* **54**, 5258–5261.
10. Chen, L., Chen, D., Manome, Y., Dong, Y., Fine, H. A., Kufe, D. W. (1995) Breast cancer selective gene expression and therapy mediated by recombinant adenoviruses containing the DF3/MUC1 promoter. *J. Clin. Invest.* **96**, 2775–2782.
11. Parr, M. J., Manome, Y., Tanaka, T., Wen, P., Kufe, D. W., Kaelin, W. G., Jr., Fine, H. A. (1997) Tumor-selective transgene expression in vivo mediated by an E2F-responsive adenoviral vector. *Nat. Med.* **3**, 1145–1149.
12. Kyo, S., Inoue, M. (2002) Complex regulatory mechanisms of telomerase activity in normal and cancer cells: how can we apply them for cancer therapy? *Oncogene* **21**, 688–697.
13. Takakura, M., Kyo, S., Kanaya, T., Hirano, H., Takeda, J., Yutsudo, M., Inoue, M. (1999) Cloning of human telomerase catalytic subunit (hTERT) gene promoter and identification of proximal core promoter sequences essential for transcriptional activation in immortalized and cancer cells. *Cancer Res.* **59**, 551–557.
14. Huang, H.-S., Chiou, J.-F., Fong, Y., Hou, C.-C., Lu, Y.-C., Wang, J.-Y., Shih, J.-W., Pan, Y.-R., Lin, J.-J. (2003) Activation of human telomerase reverse transcriptase expression by some new symmetrical bis-substituted derivatives of the anthraquinone. *J. Med. Chem.* **46**, 3300–3307.
15. Lin, S.-C., Li, W.-C., Shih, J.-W., Hong, K.-F., Pan, Y.-R., Lin, J.-J. (2006) The tea polyphenols EGCG and EGC repress mRNA expression of human telomerase reverse transcriptase (hTERT) in carcinoma cells. *Cancer Lett.* **236**, 80–88.
16. Huang, H.-S., Chou, C.-L., Guo, C.-L., Yuan, C.-L., Lu, Y.-C., Shieh, F.-Y., Lin, J.-J. (2005) Human telomerase inhibition and cytotoxicity of regioisomeric disubstituted amidoanthraquinones and aminoanthraquinones. *Bioorg. Med. Chem.* **13**, 1435–1444.
17. Denizot, F., Lang, R. (1986) Rapid colorimetric assay for cell growth and survival. Modifications to the tetrazolium dye procedure giving improved sensitivity and reliability. *J. Immunol. Methods* **89**, 271–277.

9

Telomerase RNA Inhibition Using Antisense Oligonucleotide Against Human Telomerase RNA Linked to a 2′,5′-Oligoadenylate

Yasuko Kondo and Seiji Kondo

Summary

Telomerase, a ribonucleoprotein enzyme, is detected in the vast majority of cancers, including malignant gliomas, but not in most normal somatic cells. To inhibit telomerase function effectively, we have adopted the 2′,5′-oligoadenylate (2-5A) antisense system. 2-5A is a mediator of one pathway of interferon actions by activating RNase L, resulting in single-stranded RNA cleavage. By linking 2-5A to an antisense oligonucleotide, RNase L degrades the targeted RNA specifically and effectively. Therefore, we have synthesized the antisense oligonucleotide against human telomerase RNA component (hTR) linked to 2-5A (2-5A-anti-hTR) and have demonstrated its antitumor effect on telomerase-positive cancer cells in vitro and in vivo.

Key Words: Cancer; telomerase; telomerase RNA; 2-5A; antisense; apoptosis.

1. Introduction
1.1. Telomerase Inhibition as Anticancer Therapy

Telomerase, a ribonucleic acid–protein complex, adds hexameric repeats of 5′-TTAGGG-3′ to the ends of telomeres to compensate for the progressive loss of telomeres that occurs over the course of DNA replication *(1–3)*. Telomerase activity, which is generally undetectable in normal somatic cells, is seen in approximately 90% of tumors *(4,5)*. Therefore, telomerase is expected to be a

From: *Methods in Molecular Biology, vol. 405: Telomerase Inhibition*
Edited by: L. G. Andrews and T. O. Tollefsbol © Humana Press Inc., Totowa, NJ

very good candidate as not only a useful prognostic and diagnostic marker but also a targeted therapy for various cancers, including malignant gliomas *(6,7)*.

Recently, three major subunits of the human telomerase complex have been identified: human telomerase RNA component (hTR) *(8)*, telomerase-associated protein 1 (TEP1) *(9,10)*, and human telomerase reverse transcriptase (hTERT) *(11,12)*. hTR functions as a template for telomere elongation by telomerase. TEP1, which is homologous to the gene of *Tetrahymena* telomerase component p80, is associated with RNA and protein binding. hTERT contains reverse transcriptase motifs and functions as the catalytic subunit of telomerase.

Strategies for the inhibition of telomerase have focused on antisense against hTR or hTERT, inhibitors of reverse transcriptases, small molecules able to interact with and stabilize four-stranded (G-quadruplex) structures formed by telomeric DNA, and telomerase template antagonists such as peptide nucleic acid and 2′-*O*-MeRNA oligomers *(13–17)*. These investigations have shown that inhibiting telomerase suppresses the proliferation of cancer cells and the growth of tumor xenografts in animals, making telomerase a promising target in cancer therapy. However, one limitation of targeting telomerase in tumors is that telomeric DNA must be shortened considerably before the cells undergo cell death or senescence. This means there will be a lag phase between the time telomerase is inhibited and the time telomeres of the cancer cells will have shortened sufficiently to have a significant effect on cell proliferation. Therefore, the lag time needs to be shortened by modification of the approach to telomerase inhibition or exploration of a telomerase inhibition system that induces rapid cell death.

1.2. 2′,5′-Oligoadenylate-Anti-hTR Therapy for Telomerase-Positive Cancer Cells

The 2′,5′-oligoadenylate (2-5A)-antisense system is a novel technology that exploits the body's natural antiviral defense by recruiting RNase L *(18,19)*, an endoribonuclease that functions in the interferon-regulated 2-5A system to degrade viral and cellular single-stranded RNAs. It is converted from a silent form to an active form upon binding to 2-5A. As shown in **Fig. 1**, to activate RNase L and direct it to a specific RNA target, 2-5A is attached through linkers to the 5′ terminus of an antisense oligonucleotide, creating 2-5A antisense. This system has been targeted against many RNAs involved in human disease, including respiratory syncytial virus mRNA *(19)* and *bcl/abl* mRNA in chronic myelogenous leukemia *(20)*. Because RNase L is present in most mammalian cells, the application of this technology is a powerful method for controlling gene expression. Therefore, as shown in **Fig. 2**, we decided

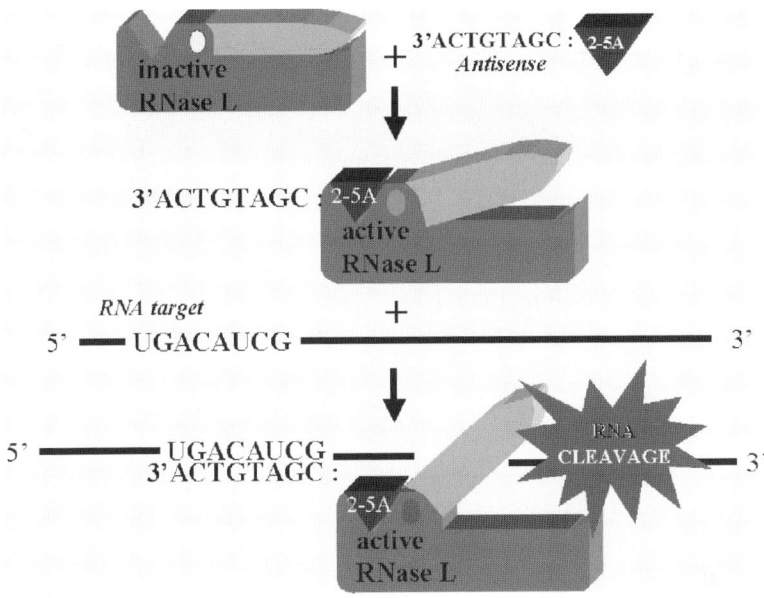

Fig. 1. Targeted degradation of RNA by 2′,5′-oligoadenylate (2-5A) antisense chimeras. RNase L is an endoribonuclease to degrade viral and cellular single-stranded RNAs. It is converted from a silent to an active form upon binding to 2-5A. To activate RNase L and direct it to a specific RNA target, 2-5A is attached through linkers to the 5′ terminus of an antisense oligonucleotide, creating 2-5A antisense. (Modification of **ref.** *(25)*.)

to use 2-5A antisense technology to degrade hTR specifically and effectively *(21)*. As we and others have previously demonstrated, treatment with 2-5A-anti-hTR degrades hTR, inhibits telomerase activity, and consequently induces apoptosis in malignant glioma, prostate cancer, bladder cancer, cervical cancer, and ovarian cancer cells *(21–26)*. The cell death is caspase-dependent and telomere length-independent *(23,24,26)*. Although the molecular mechanisms by which 2-5A-anti-hTR induces telomere length-independent cell death are not fully understood, it is possible that telomere-capping effect by telomerase *(27)* is relieved by the treatment with 2-5A-anti-hTR, which may stimulate growth-inhibiting signals and lead to immediate cell death. Moreover, the intratumoral injections of 2-5A-anti-hTR are effective for subcutaneous or intracranial tumors in nude mice *(21–24,26)*. The in vitro and in vivo effects of 2-5A-anti-hTR are significantly enhanced by its combination with cisplatin or Ad5CMV-p53, a recombinant adenovirus carrying the tumor-suppressor

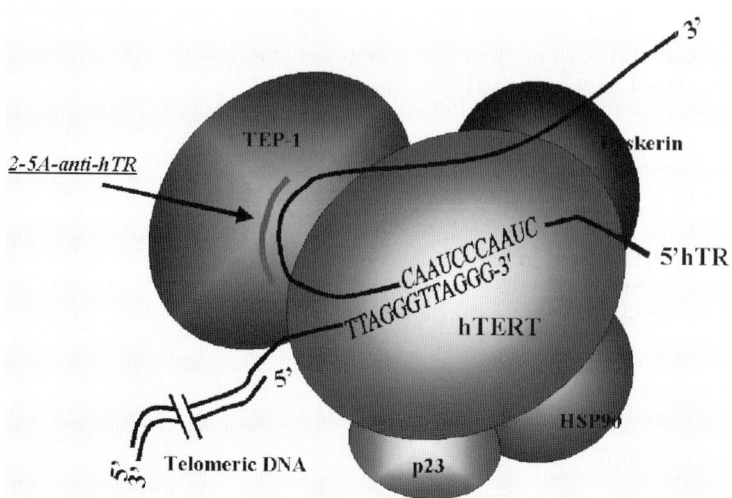

Fig. 2. Telomere/telomerase complex and target of 2′,5′-oligoadenylate-anti-human telomerase RNA component (2-5A-anti-hTR). Recently, six subunits of the human telomerase complex have been identified: hTR *(8)*, telomerase-associated protein 1 (TEP1) *(9,10)*, human telomerase reverse transcriptase (hTERT) *(11,12)*, heat-shock protein 90 (HSP90) *(37)*, molecular chaperon p23 *(37)*, and dyskerin *(38)*. Therefore, we decided to use 2-5A antisense technology to degrade hTR specifically and effectively *(21)*.

p53 *(28,29)*. These findings indicate that targeting telomerase RNA with 2-5A-anti-hTR may be an effective strategy for the management of various cancers with telomerase activity.

2. Materials

2.1. Synthesis of 2-5A-Anti-hTR

The selection of the antisense against hTR has been described previously *(21)*. Briefly, we designed the 2-5A antisense oligonucleotide to the hTR region between residues 76 and 94 because the mfold computer program (Zucker and Jaeger, Ottawa, Ontario, Canada) predicted that sequence would be the most open. As summarized in **Table 1**, the sequence of the antisense is 5′-GCGCGGGGAGCAAAAGCAC-3′. To investigate the effect of 2-5A-anti-hTR, we synthesized a test oligonucleotide (spA_4-anti-hTR) with complete homology to the targeted sequence and two control oligonucleotides. One control oligonucleotide was spA_2-anti-hTR, which had a non-functional chimeric 2-5A linked to the antisense hTR. The other control oligonucleotide was spA_4-anti-(M6)hTR,

Table 1
Nomenclature and Sequence of 2′,5′-Oligoadenylate-Anti-Human Telomerase RNA Component (2-5A-Anti-hTR)

Oligonucleotide	Sequence
spA$_4$-anti-hTR	sp5′A(2′p5′A)$_3$-Bu$_2$-5′GCGCGGGGAGCAAAAGCAC3′-3′T5′
spA$_2$-anti-hTR	sp5′A2′p5′A-Bu$_2$-5′GCGCGGGGAGCAAAAGCAC3′-3′T5′
spA$_4$-anti-(M6)hTR	sp5′A(2′p5′A)$_3$-Bu$_2$-5′GCCCGCGGTGCTAATGCTC3′-3′T5′

Bu indicate butanediol linkers; underline indicates mismatched nucleotides.

which contained functional 2-5A but with six mismatched nucleotides in the antisense that would prevent homologous binding with the telomerase RNA. These chimeric oligonucleotides were synthesized on solid supports and purified as described previously *(18,19)*. Currently, 2-5A-anti-hTR is synthesized at Sigma-Genosys (Woodlands, TX), with a modification of spacer 19 instead of a butanediol linker to link 2-5A to the oligonucleotide. This modified 2-5A-anti-hTR has an antitumor effect similar to that of the original one. The stock solution of 100 μM is prepared with sterile distilled water and kept at -20 °C. 2-5A-anti-hTR keeps the cytotoxic activity during at least 10× freeze-thaw cycles and even after 1-week incubation at 37 °C.

2.2. Cell Culture

1. Human malignant glioma cells are purchased from American Type Culture Collection (Manassas, VA).
2. Cells are cultured in Dulbecco's modified Eagle's medium (Cellgro, Herndon, VA) supplemented with 10% fetal bovine serum (Invitrogen, Carlsbad, CA), 100 U/ml penicillin (Invitrogen), and 2.5 μg/ml fungizone (Invitrogen) at 37°C in 5% CO$_2$.
3. Cells are trypsinized by 0.05% trypsin–ethylenediamine tetraacetic acid (EDTA) (Invitrogen).
4. Phosphate-buffered saline (PBS) (Mg^{2+}-free and Ca^{2+}-free; Cellgro).

2.3. Cell Viability Assay

The cytotoxic effect of 2-5A-anti-hTR on tumor cells is determined by using a cell-proliferation reagent *(30)*.

1. Cell-proliferation reagent WST-1 (Roche Applied Science, Indianapolis, IN).
2. Lipofectamine (Invitrogen).
3. 96-well flat-bottomed plates (Becton Dickinson and Company, Franklin Lakes, NJ).
4. Microplate reader (EL808, Ultra Microplate Reader; BIO-TEK, Winooski, VT).

2.4. Reverse-Transcription Polymerase Chain Reaction for hTR

To determine whether hTR is degraded by 2-5A-anti-hTR, the expression of hTR is assessed by reverse-transcription polymerase chain reaction (RT-PCR) *(21)*.

1. RNeasy Mini Kit (Qiagen, Valencia, CA).
2. RNase-free water (Qiagen).
3. 20-gauge or 22-gauge needles with 1-ml to 5-ml syringes (Fisher, Newark, DE).
4. ThermoScript RT-PCR System (Invitrogen).
5. Thermocycler (MJ Research/Bio-Rad, Hercules, CA).
6. The primer sets used are as follows: glyceraldehyde 3-phosphate dehydrogenase (GAPDH), 5'-CTCAGACACCATGGGGAAGGTGA-3' (forward) and 5'-ATGATCTTGAGGCTGTTGTCATA-3' (reverse); hTR, 5'-TCTACCCTAACTGAGAAGGGCGTAG-3' (forward) and 5'-GTTTGCTCTAGAATGAACGGTGGAAG-3' (reverse).
7. 2% agarose gel containing 0.5 μg/ml ethidium bromide.
8. CCD Video Camera/Imaging System IS200M-110 (Alpha Innotech, San Leandro, CA).

2.5. Telomere Repeat Amplification Protocol Assay

This assay measures enzymatic activity of telomerase in tumor cells treated with 2-5A-anti-hTR by using a TRAPEZE telomerase detection kit (Chemicon International, Temecula, CA) *(31)*.

1. TRAPEZE telomerase detection kit (S7700).
2. Thermocycler (MJ Research/Bio-Rad).
3. Polyacrylamide vertical gel electrophoresis (PAGE) apparatus (Bio-Rad).
4. Power supply (>500 V capacity) (Bio-Rad).
5. Taq DNA polymerase (Invitrogen).
6. PBS (Mg^{2+}-free and Ca^{2+}-free; Cellgro).
9. Reagents for PAGE: 40% polyacrylamide/bisacrylamide stock solution (19:1) (MD Biochemicals, Aurora, OH), *N,N,N,N*'-tetramethyl-ethylenediamine (TEMED; Bio-Rad), 10% ammonium persulfate (MD Biochemicals), 10× TBE solution (1 l requires 54 g Tris base, 27.5 g boric acid, and 20 ml 0.5 M EDTA, pH 8.0). SYBR Green (Molecular Probes, Eugene, OR). UV box: 254 or 302 nm for SYBR Green.
10. CCD Video Camera/Imaging System IS200M-110 (Alpha Innotech).

2.6. Apoptosis Detection Assay

To determine whether 2-5A-anti-hTR induces apoptosis in tumor cells, the ApopTag peroxidase in situ apoptosis detection kit (cat. no. S7100, Chemicon International) is used.

1. Deionized water (dH$_2$O).
2. 2-well Lab-Tex chamber slide (Nalge Nunc International, Rochester, NY).
3. 4% paraformaldehyde (Sigma Chemical Co., St. Louis, MO).
4. Slide mounting medium (Permount; Fisher).
5. AX70 microscopy (Olympus, Melville, NY).

2.7. In Vivo Antitumor Effect of 2-5A-Anti-hTR

2.7.1. Effect of 2-5A-Anti-hTR on Subcutaneous Tumors

Before treatment of intracranial tumors in nude mice with 2-5A-anti-hTR, we optimized the in vivo ratio of 2-5A-anti-hTR to lipofectamine using subcutaneous tumors *(21–24,26)*.

1. Nude mice (8–12 weeks old; 20–25 g weight; Jackson Laboratory, Bar Harbor, MA).
2. 27-gauge tuberculin syringe (Becton Dickinson and Company).
3. Optimal cutting temperature (OCT) compound and glass slides (VWR, West Chester, PA).

2.7.2. Effect of 2-5A-Anti-hTR on Intracranial Tumors

On the basis of the results of treating subcutaneous tumors with 2-5A-anti-hTR, intracranial tumors are treated. To enhance the delivery of 2-5A-anti-hTR into the intracranial tumors, convection-enhanced delivery (CED) *(32)* is used. In CED, therapeutic agents are delivered in continuous small pulses through an intratumoral catheter, resulting in convective distribution of those agents within the tumor. Moreover, CED can circumvent the blood-brain barrier and minimize systemic toxic effects of therapy. Evidence is increasing that chemotherapeutic agents, toxins, monoclonal antibodies, and boronated epidermal growth factors can be efficiently delivered to large regions of the brain with CED *(32–35)*.

1. Nude mice (8–12 weeks old; 20–25 g weight; Jackson Laboratory). Mice are anesthetized with intraperitoneal injection of ketamine (100 mg/kg) and xylazine (5–16 mg/kg) *(22)*.
2. A screw guide with a 0.5-mm central hole and a stylet (Plastics One, Roanoke, VA).
3. A small hand-controlled twist drill with a bur 1 mm in diameter (Plastics One).
4. A 10-μl Hamilton syringe fitted with a 26-gauge needle (Fisher).
5. A microinfusion pump Harvard PHD 2000 Injector (Harvard Apparatus, Holliston, MA).

3. Methods

3.1. Cell Viability Assay

1. Seed the cells at 5×10^3 cells per well (0.1 ml) in 96-well flat-bottomed plates and incubate them overnight at 37 °C.

2. Remove the culture medium and add 2-5A-anti-hTR (5.0 μM) with 0.4 μl lipofectamine in 10 μl sterile distilled water to the wells. Incubate the plates for 1 min and add 90 μl fresh medium. The final concentration of 2-5A-anti-hTR is 0.5 μM. As a control, use lipofectamine (0.4 μl) alone in 10 μl sterile distilled water (*see* **Note 1**).
3. Treat the cells with 2-5A-anti-hTR every 24 h until the WST-1 assay is done. Usually, 3–4 days are needed to inhibit the proliferation of cancer cells to less than 50% of the control.
4. Mix 400 μl WST-1 with 3600 μl media for 40 wells. After the old medium is removed from each well, apply 100 μl WST-1 mixed solution to each well. Then, incubate the cells for 1 h at 37 °C and analyze using a microplate reader with a filter for a wavelength between 420 and 480 nm. The reagents for WST-1 are included in the kit.
5. Calculate the survival fractions from the mean cell viability of treated cells divided by that of cells with control treatment (lipofectamine alone).

3.2. RT-PCR for hTR

3.2.1. Isolation of Total RNA (RNeasy Mini Kit, Including Necessary Reagents)

1. Mix 600 μl buffer RLT and 6 μl β-mercaptoethanol for each sample. Harvest treated cells with trypsinization and wash with PBS. There should be $<1 \times 10^7$ cells/sample. Add 600 μl buffer RLT including β-mercaptoethanol and mix well. For homogenization, pass the lysate $\geq 5\times$ through a 20-gauge or 22-gauge needle syringe.
2. Add 600 μl 70% ethanol (dilute ethanol with RNase-free water) and mix well. Apply 700 μl sample to a RNeasy mini-column placed in a 2-ml collection tube. Centrifuge for 15 s at $9,300 \times g$ and discard the flow-through. Apply another sample (about 500 μl) to the same column. Centrifuge for 15 s at $9,300 \times g$ and discard the flow-through.
3. Add 700 μl buffer RW1 to the column. Centrifuge for 15 s at $9,300 \times g$ and discard both the flow-through and collection tube. Set the column on a new 2-ml collection tube.
4. Add 500 μl buffer RPE to the column. Centrifuge for 15 s at $9,300 \times g$ and discard the flow-through. Add another 500 μl buffer RPE to the column. Centrifuge for 2 min at $9,300 \times g$ and discard the flow-through. Centrifuge for 1 min at $16,900 \times g$ and discard both the flow-through and collection tube.
5. Set the column on a new 1.5-ml collection tube. Add 30–50 μl RNase-free water to the column. Centrifuge for 1 min at $9,300 \times g$ and discard the column. Measure concentration or store at −20 or −80 °C.

3.2.2. Measurement of Concentration of RNA

1. Dilute the RNA with dH$_2$O at an appropriate rate. (Recommendation: $\times 250$, 2 μl RNA +498 μl H$_2$O.) Prepare 500 μl dH$_2$O for a blank.

2. After calibration with the blank, measure the absorbance at 260 nm (A_{260}) with a spectrophotometer. After measurement, discard the diluted RNA.
3. Calculate the concentration of RNA. Concentration of RNA (μg/ml) $= 40 \times A_{260} \times$ dilution factor. If 1:250 dilution, concentration (μg/μl) $= 10 \times A_{260}$.

3.2.3. RT (Preparation of cDNA; ThermoScript RT-PCR System, Including Necessary Reagents)

1. Mix 1 μg RNA, 2 μl 10 mM dNTP mix, and 1 μl primer (usually, Random Hexamers) in RNase-free water to yield 12 μl for each sample.
2. Incubate at 65°C for 5 min and then place it on ice. Mix the 5× cDNA synthesis buffer well.
3. Prepare cDNA synthesis mix: 4 μl 5× cDNA synthesis buffer, 1 μl dithiothreitol (DTT) (0.1 M), 1 μl RNase-free water, 1 μl RNase OUT, and 1 μl ThermoScript RT to yield 8 μl for each sample.
4. Add 8 μl cDNA synthesis mix to each reaction. Put reactions into a thermal cycler. Set the program (25°C for 10 min, 50°C for 50 min, 85°C for 5 min, and 4°C for 99 h, 59 min, 59 s) and start. (Option) Add 1 μl RNase H and incubate at 37°C for 20 min. Use for PCR or store at −20°C.

3.2.4. PCR (ThermoScript RT-PCR System)

1. Prepare PCR mixture without Taq DNA polymerase: 5 μl 10× PCR buffer minus Mg, 1.5 μl 50 mM MgCl$_2$, 1 μl 10 mM dNTP mix, 1 μl 10 μM sense primer, 1 μl 10 μM antisense primer, and 38.1 μl dH$_2$O to yield 47.6 μl.
2. Add 0.4 μl/reaction (4 μl/10 reactions) of Taq DNA polymerase and mix well. Mix 48 μl PCR mixture with Taq and 2 μl cDNA in a PCR tube.
3. Put reactions into a thermal cycler. Set the program and start. The thermal cycles are 94°C for 1 min, 60°C for 2 min, and 70°C for 2 min for 30 cycles for GAPDH (450 bp) and 94°C for 1 min, 60°C for 1 min, and 72°C for 1 min for 30 cycles for hTR (126 bp) (*see* **Note 2**).
4. Analyze 8 μl (or an appropriate amount) amplified reactions by 2% agarose gel electrophoresis.

3.3. Telomere Repeat Amplification Protocol Assay

3.3.1. Sample Preparation

1. After tumor cells are treated with 2-5A-anti-hTR, pellet the treated cells, wash once with PBS, repellet, and carefully remove all PBS. Then, the cell pellets can be stored at −85 to −75°C or kept on dry ice (*see* **Note 3**). When thawed for extraction, resuspend the cells in 1× CHAPS lysis buffer (*see* **Note 4**).
2. Resuspend the cell pellets in 200 μl 1× CHAPS lysis buffer/10^6 cells. Incubate the suspension on ice for 30 min and centrifuge the sample at 12, 000 × g for 20 min at 4°C.

3. Determine the protein concentration of the supernatant. Aliquot the lysate in Eppendorf tubes and store at -85 to $-75\,°C$. Dilute the stock aliquot 1:20 with $1\times$ CHAPS lysis buffer before use (*see* **Note 5**).

3.3.2. Assay Set-Up

1. The reagents for telomere repeat amplification protocol (TRAP) assay are included in the kit. Thaw all reagents and store them on ice. The amount of reagents required in each assay is $5\,\mu l$ $10\times$ TRAP reaction buffer, $1\,\mu l$ $50\times$ dNTP mix, $1\,\mu l$ TS primer, $1\,\mu l$ TRAP primer mix, $0.4\,\mu l$ Taq DNA polymerase (5 units/μl), $39.6\,\mu l$ dH_2O, and $2\,\mu l$ cell extract for a total volume of $50\,\mu l$. Heat-inactivated cell extract at 85°C for 10 min is used as a negative control.

3.3.3. PCR

1. Place the tubes in the thermocycler block and incubate at $30\,°C$ for 30 min. Perform two-step PCR at $94\,°C$ for 30 s and $59\,°C$ for 30 s for 30–33 cycles in a thermocycler.

3.3.4. PAGE and Data Analysis

1. Add $10\,\mu l$ loading dye containing bromophenol blue and xylene cyanol (0.25% each in 50% glycerol/50 mM EDTA) into each reaction tube.
2. Load and run $25\,\mu l$ this on a 10% non-denaturing PAGE (no urea) in $0.5\times$ TBE buffer using PROTEAN II xi Vertical Electrophoresis Cells (Bio-Rad) (*see* **Note 6**). Run time: 1.5 h at 400 V for a 10-cm to 12-cm vertical gel.
3. After electrophoresis, stain the gel with SYBR Green at 1:10,000 in dH_2O or $0.5\times$ TBE. Stain for 30 min at room temperature and record the result by UV transilluminator (*see* **Note 7**).
4. For untreated telomerase-positive tumor cells, a ladder of products with six base increments starting at 50 nucleotides and a 36-bp internal control band are seen. In tumor cells treated with 2-5A-anti-hTR, the intensity of a ladder pattern can be decreased. In telomerase-negative samples, only the 36-bp internal control band is observed. Quantification of the telomerase activity is done by measuring the ratio of the entire ladder to the internal control band with densitometric evaluation using the CCD Video Camera/Imaging System.

3.4. Apoptosis Detection Assay

1. After tumor cells seeded into the chamber slides are treated with 2-5A-anti-hTR, fix treated cells using 4% paraformaldehyde (in case of tissue, use 95% ethanol for 15 min). Wash the samples $2\times$ for 5 min each in PBS. The reagents for apoptosis detection are included in the kit. Quench endogenous peroxidase using 1% H_2O_2 for 30 min. Rinse samples $2\times$ for 5 min each in PBS.

2. While samples are rinsed, make working-strength terminal deoxynucleotidyl transferase (TdT) enzyme: $77\,\mu l$ reaction buffer and $33\,\mu l$ TdT enzyme to yield $110\,\mu l$ working-strength TdT enzyme.

3. Tap off excess liquid and immediately apply $75\,\mu l/5\,cm^2$ of equilibration buffer, cover with a plastic cover slip, and incubate for $\geq 10\,s$ at room temperature.

4. Tap off excess liquid and immediately apply $55\,\mu l/5\,cm^2$ of working-strength TdT enzyme, cover with a cover slip, and incubate in a humidified chamber for 1 h at $37\,°C$.

5. While the samples are incubated, prepare the working-strength stop/wash buffer in a Coplin jar: 1 ml of stop/wash buffer and 34 ml dH_2O for a total volume of 35 ml.

6. After incubation, remove the plastic cover slips and place the samples in the working-strength stop/wash buffer, agitate it for 15 s, and then incubate it for 10 min at room temperature.

7. While the samples are incubating, set aside $65\,\mu l/5\,cm^2$ of anti-digoxigenin peroxidase and allow it to warm to room temperature. After incubation in stop/wash buffer, wash the samples $3\times$ for 1 min each in PBS.

8. Gently remove excess liquid and apply $65\,\mu l/5\,cm^2$ of anti-digoxigenin peroxidase to the samples, cover with a plastic cover slip, and incubate the samples in a humidified chamber at room temperature for 30 min.

9. Wash the sample $4\times$ in PBS. While washing, prepare working-strength peroxidase substrate.

10. Tap off excess liquid and apply enough peroxidase substrate to completely cover the specimen ($75\,\mu l/5\,cm^2$) and stain for 3–6 min at room temperature (*see* **Note 8**).

11. Wash the specimen $3\times$ with dH_2O, mount under a glass coverslip in a slide mounting medium, and view the chamber slides under a microscope.

3.5. In Vivo Antitumor Effect of 2-5A-Anti-hTR

3.5.1. Effect of 2-5A-Anti-hTR on Subcutaneous Tumors

1. Collect the cultured tumor cells (e.g., malignant glioma cell line U87-MG) and wash $2–3\times$ with serum-free medium. Resuspend the cells in serum-free medium at 2×10^7 cells/ml in Eppendorf tubes. Keep them on ice and take them to the animal facility room.

2. Inoculate $50\,\mu l$ cell solution (1×10^6 cells) subcutaneously into the right flank of the mouse.

3. Measure the size of tumor daily. Tumor volumes are calculated using the formula $(L\times W^2)/2$, where L = length in millimeter and W = width in millimeter.

4. To simulate clinical conditions, treatment is initiated after the tumors are established. When the tumors reach a mean tumor volume of 50–70 mm³, 2-5A-anti-hTR (1.0 nmol/tumor) with lipofectamine ($0.3\,\mu l$) in $25\,\mu l$ PBS is directly injected into the tumors every 24 h for a week. The treatment groups are (i) lipofectamine alone in PBS, (ii) Lipofectamine + a test oligonucleotide (spA$_4$-anti-hTR) in PBS,

and (iii) lipofectamine + control oligonucleotides (non-functional spA$_2$-anti-hTR or mismatch spA$_4$-anti-(M6)hTR) in PBS.

5. For histologic analysis, euthanize the mice by CO$_2$ inhalation and then remove the tumors and freeze them rapidly. Embed the frozen specimens in OCT compound and slice them into 8-μm-thick to 10-μm-thick sections, mount them on glass slides, and stain them with hematoxylin and eosin or subject to a TUNEL (Terminal deoxynuclotidyl Transferase Biotin-dUTP Nick End Labeling) assay.

3.5.2. Effect of 2-5A-Anti-hTR on Intracranial Tumors

1. Perform anesthesia by intraperitoneal injection with ketamine (100 mg/kg) and xylazine (5–16 mg/kg).
2. Sterilize the skin with povidone iodine solution and then make an incision about 3-mm long just to the right of midline and anterior to the interaural line so that the coronal and sagittal sutures and the bregma can be identified. Mark the guide-screw entry site at a point 2.5-mm lateral and 1-mm anterior to the bregma. This point is chosen because it is located directly above the caudate nucleus, which has been shown to be a highly reliable intracranial site for tumor engraftment.
3. Using a small hand-controlled twist drill with a bur 1 mm in diameter, drill a hole in the animal's skull at the entry point. Care must be taken to avoid widening the hole beyond the bit diameter because subsequent placement of the guide screw requires that the hole be exactly the size of the screw. The drill bit penetrates the dura and thereby opens it.
4. Rotate the sterilized guide screw into the hole until it is flush with the skull. This maneuver is facilitated by using a specially devised screwdriver (Plastics One) that holds the guide screw. The screw is threaded into the hole and secured with several firm twists. The top of the screw is approximately 1 mm above the skull surface, and its shaft protrudes through the dura and into the brain surface.
5. Close the central hole of the guide screw by placing a cross-shaped stylet inside it. This prevents tissue from growing into the guide screw hole (*see* **Notes 9** and **10**).
6. After 3 days, collect the cultured tumor cells and wash 2–3× with serum-free medium. Resuspend the cells in serum-free medium at 5×10^7/ml in an Eppendorf tube. Keep the tube on ice and take them to the animal facility room.
7. Reanesthetize the mice as described in 3.5.2.1. Open the wound and remove the stylets with a fine forceps. Mix or vortex the cell suspension and then draw 10 μl suspension with the cuffed Hamilton syringe. Insert the needle of the Hamilton syringe at the entry point of the guide screw. Attach the mouse and Hamilton syringe to the Harvard PHD 2000 Injector. Inject the cell suspension into the mouse brain at a rate of 1.0 μl/min. Do not remove the needle immediately after the injection. To avoid the back flow, wait for 1 min before removing the needle. Reposition the stylet in the screw hole by using a fine forceps to close the system.
8. To determine when to initiate the treatment to best simulate the clinical situation, U87-MG cells expressing green fluorescent protein (GFP) were inoculated into the

brain of nude mice as described in 3.5.2.7. GFP-positive tumor cells ($\sim$1.2 mm in diameter) were detected in the cerebral hemisphere under a fluorescence microscope 3 days after tumor inoculation *(36)*. Therefore, when U87-MG cells are used, initiate the treatment with 2-5A-anti-hTR 3 days after tumor inoculation and inject 3.0 nmol 2-5A-anti-hTR/1.0 µl lipofectamine in 10 µl sterile PBS through a microinfusion pump at an infusion rate of 1.0 µl/min through the same screw guide. Repeat the treatment every other day for 10 days.

9. When animals exhibit any clinical signs of neurological toxicity (e.g., paresis), euthanize and study them immediately. Assess the significance of between-group differences in survival time using the Cox-Mantel test *(22)*. For histologic analysis, remove the brains, slice them coronally into five blocks, and snap-freeze before slicing into 8-µm to 10-µm coronal sections.

4. Notes

1. Treatment with 2-5A-anti-hTR for in vitro experiments is conducted according to the described procedure.
2. Reactions can be stored at 4 or $-20\,^\circ$C (long term).
3. Telomerase in frozen cells is stable for ≥ 1 year at -85 to $-75\,^\circ$C.
4. Telomerase will remain stable for short periods at room temperature, but try not to leave protein extracts out for any longer than necessary, and keep them on ice when they are out. Always flash-freeze extracts in liquid nitrogen before putting them away at -80 °C.
5. Cell-extract concentration is $1–10\,\mu g/\mu l$. Use $<1.5\,\mu g$ per assay.
6. To make 50 ml 10% non-denaturing polyacrylamide gel, use 49.5 ml 10% polyacrylamide (mono/bis = 19:1) stock in 0.5× TBE, 0.5 ml 10% ammonium persulfate, and 0.05 ml TEMED. Take extreme care to prevent sample carryover into adjacent wells, which may produce false-positive results.
7. SYBR Green is toxic. Carefully handle and wear double gloves.
8. To determine the optimal staining condition, monitor color development by looking at the slide under a microscope.
9. The wound may be closed with No. 4-0 Vicryl sutures, although as experience is gained, the incisions frequently become small enough that no suturing of the skin is required.
10. Keep the mice warm until they recover from anesthesia. Animals may be allowed to move freely until the time of cell implantation.

Acknowledgments

We thank Dr Takao Kanzawa for figures and Emporia F. Hollingsworth and Dr Eiji Iwado for technical assistance. We also thank David Galloway for editing the manuscript. This study was supported in part by the grants CA-088936 and CA-108558 from the National Cancer Institute (SK), by a start-up fund from The

University of Texas M. D. Anderson Cancer Center (SK), by a generous donation from the Anthony D. Bullock III Foundation (YK and SK), and by the cancer center support grant/shared resources CA-16672 from the National Cancer Institute.

References

1. Blackburn, E.H. (1991) Structure and function of telomeres. *Nature* **350**, 569–573.
2. Counter, C.M., Avilion, A.A., LeFeuvre, C.E., Stewart, N.G., Greider, C.W., Harley, C.B., Bacchetti, S. (1992) Telomere shortening associated with chromosome instability is arrested in immortal cells which express telomerase activity. *EMBO J.* **11**, 1921–1929.
3. de Lange, T. (1994) Activation of telomerase in a human tumor. *Proc. Natl. Acad. Sci. U. S. A.* **91**, 2882–2885.
4. Kim, N.W., Piatyszek, M.A., Prowse, K.R., Harley, C.B., West, M.D., Ho, P.L.C., Coviello, G.M., Wright, W.E., Weinrich, S.L., Shay, J.W. (1994) Specific association of human telomerase activity with immortal cell lines and cancer. *Science* **266**, 2011–2015.
5. Shay, J.W., Zou, Y., Hiyama, E., Wright, W.E. (2001) Telomerase and cancer. *Hum. Mol. Genet.* **10**, 677–685.
6. Shay, J.W., Wright, W.E. (2002) Telomerase: a target for cancer therapeutics. *Cancer Cell* **2**, 257–265.
7. Komata, T., Kanzawa, T., Kondo, Y., Kondo, S. (2002) Telomerase as a therapeutic target for malignant gliomas. *Oncogene* **21**, 656–663.
8. Feng, J., Funk, W.D., Wang, S.S., Weinrich, S.L., Avilion, A.A., Chiu, C.P., Adams, R.R., Chang, E., Allsopp, R.C., Yu, J., Le, S., West, M.D., Harley, C.B., Andrews, W.H., Greider, C.W., Villeponteau B. (1995) The RNA component of human telomerase. *Science* **269**, 1236–1241.
9. Harrington, L., McPhail, T., Mar, V., Zhou, W., Oulton, R., Bass, M.B., Arruda, I., Robinson, M.O. (1997) A mammalian telomerase-associated protein. *Science* **275**, 973–977.
10. Nakayama, J., Saito, M., Nakamura, H., Matsuura, A., Ishikawa, F. (1997) *TLP1*: a gene encoding a protein component of mammalian telomerase is a novel member of WD repeats family. *Cell* **88**, 875–884.
11. Meyerson, M., Counter, C.M., Eaton, E.N., Ellisen, L.W., Steiner, P., Caddle, S.D., Ziaugra, L., Beijersbergen, R.L., Davidpoff, M.L., Liu, Q., Bacchetti, S., Haber, D.A., Weinberg, R.A. (1997) *hEST2*, the putative human telomerase catalytic subunit gene, is up-regulated in tumor cells and during immortalization. *Cell* **90**, 785–795.
12. Nakamura, T.M., Morin, G.B., Chapman, K.B., Weinrich, S.L., Andrews, W.H., Lingner, J., Harley, C.B., Cech, T.R. (1997) Telomerase catalytic subunit homologs from fission yeast and human. *Science* **277**, 955–959.
13. Kondo, S., Tanaka, Y., Kondo, Y., Hitomi, M., Barnett, G.H., Ishizaka, Y., Liu, J., Haqqi, T., Nishiyama, A., Villeponteau, B., Cowell, J.K., Barna, B.P. (1998)

Antisense telomerase treatment: induction of two distinct pathways, apoptosis and differentiation. *FASEB J.* **12**, 801–811.

14. Hahn, W.C., Stewart, S.A., Brooks, M.W., York, S.G., Eaton, E., Kurachi, A., Beijersbergen, R.L., Knoll, J.H.M., Meyerson, M., Weinberg, R.A. (1999) Inhibition of telomerase limits the growth of human cancer cells. *Nat. Med.* **5**, 1164–1170.

15. Zhang, X., Mar, V., Zhou, W., Harrington, L., Robinson, M.O. (1999) Telomere shortening and apoptosis in telomerase-inhibited human tumor cells. *Gene Dev.* **13**, 2388–2399.

16. Mergny, J.L., Helene, C. (1998) G-quadruplex DNA: a target for drug design. *Nat. Med.* **4**, 1366–1367.

17. Herbert, B., Pitts, A.E., Baker, S.I., Hamilton, S.E., Wright, W.E., Shay, J.W., Corey, D.R. (1999) Inhibition of human telomerase in immortal human cells leads to progressive telomere shortening and cell death. *Proc. Natl. Acad. Sci. U. S. A.* **96**, 14276–14281.

18. Maran, A., Maitra, R.K., Kumar, A., Dong, B., Xiao, W., Li, G., Williams, B.R.G., Torrence, P.F., Silverman, R.H. (1994) Blockage of NF-κB signaling by selective ablation of an mRNA target by 2-5A antisense chimeras. *Science* **265**, 789–792.

19. Cirino, N.M., Li, G., Xiao, W., Torrence, P.F., Silverman, R.H. (1997) Targeting RNA decay in respiratory syncytial virus infected cells with 2′->5′ oligoadenylate-antisense. *Proc. Natl. Acad. Sci. U. S. A.* **94**, 1937–1942.

20. Maran, A., Waller, C.F., Paranjape, J.M., Li, G., Xiao, W., Zhang, K., Kalaycio, M.E., Maitra, R.K., Lichtin, A.E., Brugger, W., Torrence, P.F., Silverman, R.H. (1998) 2′,5′-Oligoadenylate-antisense chimeras cause RNase L to selectively degrade bcr/abl mRNA in chronic myelogenous leukemia cells. *Blood* **92**, 4336–4343.

21. Kondo, S., Kondo, Y., Li, G., Silverman, R.H., Cowell, J.K. (1998) Targeted therapy of human malignant glioma in a mouse model by 2-5A antisense directed against telomerase RNA. *Oncogene* **16**, 3323–3330.

22. Mukai, S., Kondo, Y., Koga, S., Komata, T., Barna, B.P., Kondo, S. (2000) 2-5A antisense telomerase RNA therapy for intracranial malignant gliomas. *Cancer Res.* **60**, 4461–4467.

23. Koga, S., Kondo, Y., Komata, T., Kondo, S. (2000) Treatment of bladder cancer cells in vitro and in vivo with 2-5A antisense telomerase RNA. *Gene Ther.* **8**, 654–658.

24. Kondo, Y., Koga, S., Komata, T., Kondo, S. (2000) Treatment of prostate cancer in vitro and in vivo with 2-5A-anti-telomerase RNA component. *Oncogene* **19**, 2205–2211.

25. Kushner, D.M., Paranjape, J.M., Bandyopadhyay, B., Cramer, H., Leaman, D.W., Kennedy, A.W., Silverman, R.H., Cowell, J.K. (2000) 2-5A antisense directed against telomerase RNA produces apoptosis in ovarian cancer cells. *Gynecol. Oncol.* **76**, 183–192.

26. Yatabe, N., Kyo, S., Kondo, S., Kanaya, T., Wang, Z., Maida, Y., Takakura, M., Nakamura, M., Tanaka, M., Inoue, M. (2002) 2-5A antisense therapy directed against human telomerase RNA inhibits telomerase activity and induces apoptosis without telomere impairment in cervical cancer cells. *Cancer Gene Ther.* **9**, 624–630.

27. Blackburn, E.H., Chan, S., Chang, J., Fulton, T.B., Krauskopf, A., McEachern, M., Prescott, J., Roy, J., Smith, C., Wang, H. (2000) Molecular manifestations and molecular determinants of telomere capping. *Cold Spring Harb. Symp. Quant. Biol.* **65**, 253–263.

28. Komata, T., Kondo, Y., Koga, S., Ko, S.C., Chung, L.W.K., Kondo, S. (2000) Combination therapy of malignant glioma cells with 2-5A-antisense telomerase RNA and recombinant adenovirus p53. *Gene Ther.* **7**, 2071–2079.

29. Kondo, Y., Komata, T., Kondo, S. (2001) Combination therapy of 2-5A antisense against telomerase RNA and cisplatin for malignant gliomas. *Int. J. Oncol.* **18**, 1287–1292.

30. Ito, H., Kanzawa, T., Miyoshi, T., Hirohata, S., Kyo, S., Iwamaru, A., Aoki, H., Kondo, Y., Kondo, S. (2005) Therapeutic efficacy of PUMA for malignant glioma cells regardless of p53 status. *Hum. Gene Ther.* **16**, 685–698.

31. Kanzawa, T., Germano, I.M., Kondo, Y., Ito, H., Kyo, S., Kondo, S. (2003) Inhibition of telomerase activity in malignant glioma cells correlates with their sensitivity to temozolomide. *Br. J. Cancer* **89**, 922–929.

32. Bobo, R.H., Laske, D.W., Akbasak, A., Morrison, P.F., Dedrick, R.L., Oldfield, E.H. (1994) Convection-enhanced delivery of macromolecules in the brain. *Proc. Natl. Acad. Sci. U. S. A.* **91**, 2076–2080.

33. Viola, J.J., Agbaria, R., Walbridge, S., Oshiro, E.M., Johns, D.G., Kelley, J.A., Oldfield, E.H., Ram, Z. (1995) In situ cyclopentenyl cytosine infusion for the treatment of experimental brain tumors. *Cancer Res.* **55**, 1306–1309.

34. Laske, D.W., Youle, R.J., Oldfield, E.H. (1997) Tumor regression with regional distribution of the targeted toxin TF-CRM107 in patients with malignant brain tumors. *Nat. Med.* **3**, 1362–1368.

35. Yang, W., Barth, R.F., Adams, D.M., Ciesielski, M.J., Fenstermaker, R.A., Shukla, S., Tjarks, W., Caligiuri, M.A. (2002) Convection-enhanced delivery of boronated epidermal growth factor for molecular targeting of EGF receptor-positive gliomas. *Cancer Res.* **62**, 6552–6558.

36. Ito, H., Aoki, H., Kühnel, F., Kondo, Y., Kubicka, S., Wirth, T., Iwado, E., Iwamaru, A., Fujiwara, K., Hess, K.R., Lang, F.F., Sawaya, R., Kondo, S. (2006) Autophagic cell death of malignant glioma cells induced by a conditionally replicating adenovirus. *J. Natl. Cancer Inst.* **98**, 625–636.

37. Holt, S.E., Aisner, D.L., Baur, J., Tesmer, V.M., Dy, M., Ouellette, M., Trager, J.B., Morin, G.B., Toft, D.O., Shay, J.W., Wright, W.E., White, M.A. (1999) Functional requirement of p23 and Hsp90 in telomerase complexes. *Genes Dev.* **13**, 817–826.

38. Mitchell, J.R., Wood, E., Collins, K. (1999) A telomerase component is defective in the human disease dyskeratosis congenita. *Nature* **402**, 551–555.

10

Knockdown of Telomerase RNA Using Hammerhead Ribozymes and RNA Interference

Shang Li, Mehdi Nosrati, and Mohammed Kashani-Sabet

Summary

More than 85% of human cancers and over 70% of immortalized human cell lines have highly elevated telomerase activity. In contrast, telomerase activity is down-regulated in most human adult somatic cells, except stem cells and germ cells. These results are consistent with telomerase conferring a selective advantage for continued proliferation of malignant cells and present a unique target for cancer gene therapy. In line with this view, our recent results suggest that knockdown of telomerase RNA in human or in mouse cancer cells by ribozyme or RNA interference (RNAi) diminishes telomerase activity and inhibits cancer cell growth both in vitro and in vivo. Such telomerase inhibiting agents represent a promising novel cancer therapeutic strategy. In this chapter, we will discuss the knockdown of telomerase RNA by hammerhead ribozyme and RNAi. Both techniques are mediated by sequence-specific recognition of target RNA by a guide RNA molecule, which then results in the nucleolytic degradation of the RNA target.

Key Words: Telomerase; hTert; hTER; ribozyme; siRNA; shRNA.

1. Introduction

The ends of human chromosomes are capped by telomeres, which consist of terminal DNA repeats (TTAGGG)n bound by specific associated factors. These dynamic complexes protect chromosome ends from nucleolytic degradation, recombination, and end-to-end fusions that result in genetic instability and cell death (1,2). Telomeres are essential not only for chromosome stability but also for complete DNA replication (3). When cultured in vitro, human primary

From: *Methods in Molecular Biology, vol. 405: Telomerase Inhibition*
Edited by: L. G. Andrews and T. O. Tollefsbol © Humana Press Inc., Totowa, NJ

cells lose approximately 150–200 bp of telomeric DNA repeats from the end of linear chromosome per cell division because of the incomplete replication by DNA polymerase *(4–6)*. These primary cells eventually undergo replication senescence *(7,8)* unless the telomeres were replenished by re-activation of telomerase *(9)* or activation of alternative lengthening of telomeres (ALT) pathway *(10)*. In most eukaryotes, the telomeric repeat tracts are maintained by telomerase ribonucleoprotein complex that contains a core protein subunit (hTert) and an internal templating RNA subunit (hTER) *(11,12)*. Genetic inactivation of telomerase disrupts telomere maintenance and causes age-dependent and generation-dependent telomere shortening and accompanying genetic instability *(13–17)*. For in vitro reconstitution of telomerase activity in human cells, the catalytic protein subunit, hTert, and the RNA subunit, hTER *(18,19)*, are sufficient.

Although hTER is ubiquiously expressed, hTert and hence telomerase activity are diminished in most adult somatic cells, except stem cells and germ cells *(18–22)*. In contrast, highly elevated telomerase activity *(20,23)* can be detected in most human cancers. There is great interest in developing new therapeutic strategies to inhibit telomerase activity in cancer cells by targeting either the hTert or the hTER core subunit of telomerase. In this chapter, we will emphasize the knockdown of telomerase RNA by hammerhead ribozyme and RNA interference (RNAi). Both methods result in the nucleolytic degradation of the RNA and hence the inhibition of telomerase activity. Our recent results suggested that knockdown of telomerase RNA expression in human or in mouse cancer cells results in the inhibition of telomerase activity and triggers rapid cell growth inhibition both in vitro and in vivo that is independent of telomere length *(24–26)*. This is different from previous attempts to attenuate telomerase function in cell culture that results in telomere shortening and eventually cellular apoptosis *(27)*. Data from these previous studies suggested a long lag period proportional to the initial telomere length prior to the induction of significant growth inhibition.

2. Materials

2.1. DNA Cloning and Plasmid Purification

1. Plasmids including pHR'CMVGFPWSin18, pMD.G, and psPAX2 are gifts from Dr Didier Trono (School of Life Sciences, Swiss Institute of Technology, Lausanne, Switzerland).
2. Plasmid pTZ-U6+1 that contains the human U6 promoter is a gift from Dr John J. Rossi (Department of Molecular Biology, Beckman Research Institute of the City of Hope, Duarte, CA).

3. VR1225, CMV-Intron-Luciferase-Kan (Vical, San Diego, CA).
4. All DNA primers are synthesized by Integrated DNA Technologies (Skokie, IL) or Biosource (Foster City, CA).
   LS-S153: 5′-AAAAACTAGTAAGGTCGGGCAGGAAGAGGGC-3′
   LS-S161: 5′AAAACTGCAGAAAAATTGTCTAACCCTAACTGAGAATCTCTT GAATTCTCAGTTAGGGTTAGACGGTGTTTCGTCCTTTCCACAAG-3′
   EHB10424: 5′AAAACTGCAGAAAAATTGTTCTTGCGATTGTCTCTATCTCT TGAATAGAGACAATCGCAAGAACGGTGTTTCGTCCTTTCCACAAG-3′
   Ribozyme 180: 5′-TGCGCTCTGATGAGTCCGTGAGGACGAAACGTTTG-3′
   Ribozyme 180 mutant: 5′-TGCGCTCTCATGAGTCCGTGAGGACGAAACGTT TG-3′.
5. Pfu DNA polymerase (Stratagene, La Jolla, CA). Restriction endonucleases, T4 DNA ligase, and Klenow DNA polymerase (New England Biolabs, Beverly, MA).
6. HiSpeed Plasmid Maxi Kit (Qiagen, Valencia, CA).
7. EndoFree Plasmid Kit (Qiagen).

2.2. In Vitro Synthesized Small-Interference RNA

1. All synthetic short-inference RNA (siRNA) were purchased from Dharmacon (Lafayette, CO) and re-suspended in RNase-free ddH2O (*see* **Note 1**) at final 1 mM.
   siRNA-2: 5′-GUCUAACCCUAACUGAGAAUU-3′
           3′-UUCAGAUUGGGAUUGACUCUU-3′
   siRNA-3: 5′-GCAAACAAAAAAUGUCAGCUUU-3′
           3′-UUCGUUUGUUUUUUACAGUCGA-3′
   siRNA-control: 5′-GUUCUUGCGAUUGUCUCUAUU-3′
             3′-UUCAAGAACGCUAACAGAGAU-3′.

2.3. Cell Culture

1. Dulbecco's modified Eagle's medium (DMEM, Invitrogen, Carlsbad, CA) supplemented with 10% fetal bovine serum (FBS, Hyclone, Ogden, UT), 1× GlutaMAX-1 (Invitrogen), and 1× penicillin–streptomycin (Biosource, Rockville, MD).
2. RPMI 1640 (Invitrogen) supplemented with 5% FBS (Hyclone), 1× GlutaMAX-1 (Invitrogen), and 1× penicillin–streptomycin (Biosource).
3. Solution of trypsin (0.05%) and versene (0.02%) (Biosource).
4. Opti-MEM I medium (Invitrogen).
5. Lipofectamine 2000 (Invitrogen).
6. Polyvinylidene difluoride (PVDF) syringe-driven 0.45-μm filter unit (Millipore Corporation, Bedford, MA).
7. Polybrene (1,5-dimethyl-1,5-diazaundecamethylene polymethobromide) stock solution (8 mg/ml) in ddH2O (Sigma, St. Louis, MO).
8. All cell lines used in this study were obtained from ATCC (Manassas, VA).

2.4. RNA Extraction and Northern Hybridization

1. TRIzole reagent (Invitrogen) and RNeasy kit (Qiagen).
2. Glyoxal (Ethanedial) 40% aqueous solution (Sigma) deionized with AG 501-X8 resin (Bio-Rad, Hercules, CA) at 1:1 ratio (v/w) for 1 h at room temperature with rotation before use.
3. Sodium phosphate buffer pH 7.0 (10 mM for 2I): 1 M NaH_2PO_4 7.8 ml and 0.5 M Na_2HPO_4 24.4 ml in ddH2O, stored at room temperature.
4. Glyoxal cocktail:

Glyoxal (deionized)	3
0.5 M sodium phosphate buffer pH 7.0	0.72
DMSO (Sigma)	18
RNA sample	14.28
	36 μl

5. RNA loading dye (10×): 60% glycerol, 0.1 M sodium phosphate buffer pH 7.0, 0.2% bromophenol blue store at −80°C.
6. SSC (10×): 1.5 M NaCl and 0.15 M sodium citrate pH 7.0.
7. Micro-Bio-Spin 6 chromatography columns (Bio-Rad).
8. Pre-hybridization buffer: 0.5 M sodium phosphate buffer pH 7.2, 7% SDS, 1% BSA, 1 mM EDTA, store at room temperature (*see* **Note 5**).
9. RediPRIME II random prime labeling system (Amersham Biosciences, Piscataway, NJ).
10. [^{32}P]dCTP, 800 Ci/mmol, 10 mCi/ml (Perkin Elmer Life and Analytic Sciences, Boston, MA).
11. Low-stringency wash buffer: 0.1 M sodium phosphate buffer pH 7.2, 2% SDS, 1 mM EDTA, store at room temperature.
12. High-stringency wash buffer: 0.05 M sodium phosphate buffer pH 7.2, 2% SDS, 1 mM EDTA, store at room temperature.
13. Spectrophotometer (NanoDrop, Rockland, DE).
14. Bioanalyser (Agilent Technologies, Palo Alto, CA).

2.5. Telomeric Repeat Amplification Protocol Assay

1. TRAPese telomerase detection kit (Serologicals Corporation, Norcross, GA).
2. Thermocycler PTC-200 (Bio-Rad).
3. Sigmacote (Sigma).
4. EZ cast gasket (Fisher, Pittsburgh, PA).
5. SequaGel concentrate, SequaGel diluent, and SequaGel buffer (National Diagnostics, Atlanta, GA).

6. Gel electrophoresis apparatus (Model S3S, Owl Separation System, Portsmouth, NA).
7. $[\gamma-^{32}P]$ATP(3000 Ci/mmol, 10 mCi/ml) (Perkin Elmer Life and Analytic Sciences).
8. TBE (1×): 90 mM Tris-borate, 2 mM EDTA.
9. Phosphoimager scanner Storm 840 (Amersham Biosciences).
10. ImageQuant software (Amersham Biosciences).

2.6. Preparation of Cationic : Liposome Complexes

1. The cationic lipid N-[1-(2,3-dioleyloxy)propyl]-N,N,N-trimethylammonium chloride (DOTMA) was kindly provided by Dr Robert J. Debs (California Pacific Medical Research Institute, San Francisco, CA).
2. Rotary evaporator, Rotavapor R-114 (Buchi, Switzerland).
3. Dextrose (Baxter, Deerfield, IL).

2.7. Quantitative Real-Time PCR (TaqMan Assay)

1. DNase-I (RQ1, Promega, Madison, WI).
2. iScript (Bio-Rad).
3. 1× GeneAmp PCR Buffer II (Applied Biosystems, Foster City, CA).
4. 7.5 mM MgCl$_2$ solution (Applied Biosystems).
5. PCR grade dNTP (Invitrogen).
6. Random primers (Invitrogen).
7. RNaseOUT Recombinant RNase Inhibitor (Invitrogen).
8. Superscript II reverse transcriptase (Invitrogen).
9. AB Prism 7900 sequence detection system (Applied Biosystems).
10. 1× buffer A (Applied Biosystems).
11. 0.025 unit/µl AmpliTaq Gold DNA Polymerase (Applied Biosystems).
12. Primer Express software v1.5 (Applied Biosystems).
13. The TaqMan probe and primer sequences were as follows (synthesized by Integrated DNA Technologies, Skokie, IL).
    Mammalian *histone* control gene: forward primer, 5′-GCTTCCAGAGCGCAGC TATC-3′; reverse primer, 5′-GGCGTGCTAGCTGGATGTCT-3′; TaqMan probe, 5′-FAM-TGCTTTGCAGGAGGCAAGTGAGGC-TAMRA-3′.
    Mouse *gus* control gene: forward primer, 5′-CTCATCTGGAATTTCGCCGA-3′; reverse primer, 5′-GGCGAGTGAAGATCCCCTTC-3′; TaqMan probe, 5′-FAM-CGAACCAGTCACCGCTGAGAGTAATCG-TAMRA-3′.
    Ribozyme (forward primer targets the ribozyme constant region): forward primer, 5′-CTGATGAGTCCGTGAGGACGA-3′; reverse primer, 5′-GTAATTTG TCCTCCAGATCGCAG-3′; TaqMan probe, 5′-FAM-AGGTGGGCGGGCCAA GATAGGG-TAMRA-3′.

Mouse TER: forward primer, 5′-GTCTTTTGTTCTCCGCCCG-3′; reverse primer, 5′-CGGCGAACCTGGAGCTC-3′; TaqMan probe, 5′-FAM-CGTTCCCGAGC CTCAAAAACAAACG-TAMRA-3′.

2.8. Determination of Anti-Tumor Efficacy

1. 10% buffered formalin (Fisher).

3. Methods

3.1. Telomerase RNA Knockdown by RNAi

RNAi was named after the discovery that injection of double-stranded RNA (dsRNA) into the nematode *Caenorhabditis elegans* resulted in the specific silencing of genes homologous to the dsRNA delivered *(28)*. The RNAi phenomenon was subsequently confirmed in insects and mammals. The induction of RNAi is mediated by dsRNAs of 21–30 nucleotides in length that guide the RNA-induced silencing complex (RISC) protein complex to complementary RNA targets that are subsequently cleaved *(29)*. Our recent data suggested that human telomerase RNA (a non-messenger RNA) can be effec- tively suppressed by RNAi induced by small-hairpin RNA (shRNA) expressed from lentivral vector or in vitro synthesized double-stranded short-interfering RNA (siRNA). Use of lentivirus allows rapid analysis of the cellular phenotype produced without the need for antibiotic selection. The shRNAs and siRNAs were designed to target the template region and the P3 pseudoknot domain of human telomerase RNA given their known accessibility *(24)*. Expression of anti-hTER shRNAs or siRNAs (but not mutant shRNA or siRNA) results in telomerase suppression and rapid cell growth inhibition independent of telomere length and cellular status of p53 *(24,25)*.

3.1.1. Engineering of Lentivirus Vector for Expression of shRNA and Virus Preparation

1. Plasmid pTZ-U6+1, containing human U6 promoter, was used as template for PCR with primers (LS-S153/LS-S161 for anti-hTER shRNA) and (LS-S153/EHB10424 for mutant shRNA) using high-fidelity PCR enzyme pfu DNA polymerase. The PCR fragments were digested with *Spe*I and *Pst*I restriction enzymes and cloned into lentivector pHR′CMVGFPWSin18 to generate pHR′CMVGFPWSin18 U6-shRNA and pHR′CMVGFPWSin18 U6-shRNA-mut as shown in **Fig. 1A** and **B**.
2. Large-scale DNA preparation was performed using HiSpeed Plasmid Maxi Kit (cat. no. 12663, Qiagen) to purify pHR′CMVGFPWSin18 U6-shRNA, pHR′CMVGFPWSin18 U6-shRNA-mut, pMD.G (virus envelop plasmid), and pPAX2 (virus packaging plasmid).

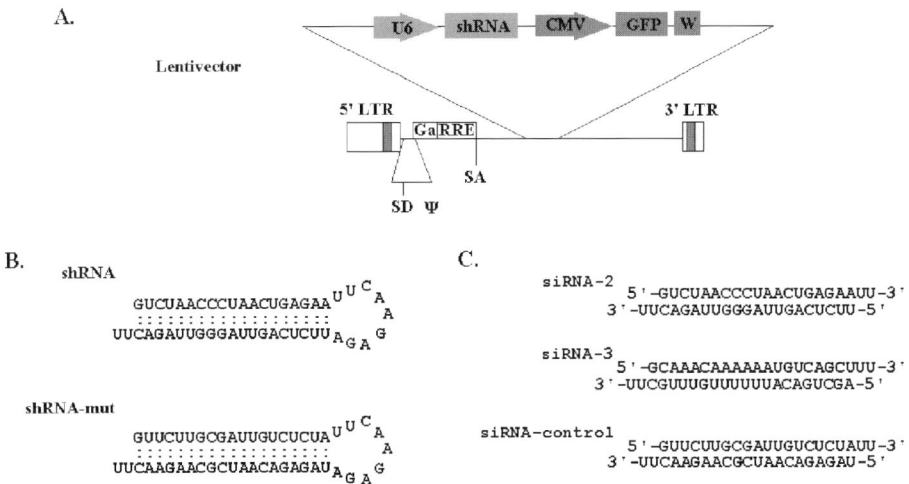

Fig. 1. Design of anti-hTER small-hairpin RNA (shRNA) expressed from lentivector and synthetic short-inference RNA (siRNA). (**A**) Lentivirus vectors: Only the relevant portion of the plasmid is illustrated. The shRNA expression cassettes inserted into the lentivector were driven by U6 promoter follow by a GFP or Puromycin reporter expression cassette. W (woodchuck hepatitis virus post-translational regulatory element). (**B**) The predicted sequence of shRNA and shRNA-mut. (**C**) The sequences for unmodified synthetic control and anti-hTER siRNAs.

3. All cell culture and virus preparation should be performed at a facility equipped with standard cell culture hood in biological safety level 2 environments. For virus production, 2×10^6 293T cells were seeded with 12 ml DMEM without antibiotics in 10-cm culture dishes overnight. The next day, about 24 μg purified DNA consisting of pHR'CMVGFPWSin18 U6-siRNA or pHR'CMVGFPWSin18 U6-shRNA-mut, pMD.G, and psPAX2 at a ratio 3:1:2 was diluted into 1.5 ml serum-free Opti-MEM I medium. In a separate tube, 60 μl lipofectamine 2000 was diluted into 1.5 ml Opti-MEM I medium and incubated at room temperature for 5 min (*see* **Note 2**). After a 5-min incubation, the DNA and lipofectamine 2000 solutions were combined, gently mixed, and incubated for additional 20 min. The lipofectamine–DNA mixture was then added to 10-cm culture dish containing 293T cells and 12 ml DMEM without antibiotics. The plate was rocked gently to mix and incubate the cells at 37°C overnight, then changed to fresh DMEM (10 ml/dish). The mediums containing virus were collected at 48 h after transfection and replaced with 10 ml fresh DMEM. The medium was then collected again at 72 h after transfection and pooled with medium previously collected. The virus-containing medium was filtered through a 0.45-μm syringe-driven filter. The flowthrough can be used immediately or aliquoted and kept at −80°C without significant loss of infectivity.

4. As the lentivirus expresses GFP marker, we can titer the virus by colony assay. Briefly, 293T cells were seeded in a 96-well culture plate at a density of 5×10^4 cells/well overnight. The lentivirus supernatant was serially diluted in DMEM and supplemented with $8 \mu g/ml$ polybrene. The diluted virus supernatant was then added to 293T cells seeded in a 96-well culture dish ($50 \mu l$/well) and incubated for 8 h before replacement with fresh DMEM. The number of GFP-positive colonies was counted under a fluorescence microscope at 48 h after virus infection. The virus titer [transduction unit (TU)/ml] = the number of GFP-positive colonies/dilution factors/50×10^{-3} ml. A typical virus supernatant has a titer of $1-5 \times 10^6$ TU/ml.

3.1.2. Cell-Culture Preparation for Treatment with shRNA-Expressing Lentivirus or Synthetic siRNA

1. For lentivirus infection, cancer cells were seeded at 5×10^5 cells/10-cm dish overnight. The next day, lentiviral supernatant at a concentration of 10–20 TU/cell (*see* **Note 3**) with $8 \mu g/ml$ polybrene in DMEM was incubated with cultured cells in 37°C incubator. About 8 h after incubation, the cells were replaced with fresh DMEM.
2. For treatment with synthetic siRNA, cancer cells were seeded at 2×10^5 cells/well in a 6-well dish. Various synthetic siRNAs as shown in **Fig. 1C** were diluted into $250 \mu l$ Opti-MEM I medium at a final of 10 nM in an Eppendorf tube. In a separate tube, $6 \mu l$ lipofectamine 2000 transfection reagent was combined with $250 \mu l$ Opti-MEM I medium and incubated for 5 min at room temperature. The siRNA solution and lipofectamine solution were combined with gentle mixing and incubated for an additional 20 min at room temperature. The siRNA–lipofectamine 2000 complex was added to each well containing cells and 2 ml DMEM. About 16 h after incubation, the cells were replaced with fresh DMEM.

3.1.3. RNA Extraction and Northern Hybridization for Detection of hTER Expression

1. For RNA extraction, $1-5 \times 10^6$ cells from each sample treated with lentivirus-expressing shRNA or synthetic siRNA as described in **Subheading 3.1.2** were harvested by trypsin, washed once with $1 \times$ PBS, and pelleted into 1.5 ml Eppendorf tube with centrifugation.
2. The cells can be stored at $-80°C$ or immediately lyzed with 1 ml TRIzole reagent in 1.5 ml RNase-free Eppendorf tube (*see* **Note 4**). The homogenized samples were incubated for 5 min at room temperature to permit the complete dissociation of nucleoprotein complex. 0.2 ml chloroform was added per 1 ml TRIzole reagent used for each sample and shaken vigorously by hand for 15 s. After 2–3 min of incubation at room temperature, the samples were centrifuged at $12,000 \times g$ for 15 min at 4°C. Following the centrifugation, the aqueous phase was transferred to a new 1.5 ml Eppendorf tube and the RNA precipitated with 0.5 ml isopropanol. Incubate sample for 10 min at room temperature and centrifuge at $12,000 \times g$ for 10 min at 4°C. Remove the supernatant and wash the RNA precipitates once with

1 ml 70% ethanol. The RNA precipitates were resuspended in RNase-free ddH2O, quantified using spectrophotometer, and stored at −80°C.

3. For RNA electrophoresis, 1.5% agarose gel was prepared with 10 mM sodium phosphate buffer in Fisher biotech gel electrophoresis system FB-SBR-2025 or similar electrophoresis system with buffer circulation apparatus. 10 μg total RNA was mixed with Glyoxal cocktail in 36 μl final volume and incubated at 50°C for 1 h. The samples were cooled to 4°C, and 4 μl 10× RNA loading dye added, and the samples loaded on an agarose gel. The gel is run in 10 mM sodium phosphate buffer at 120–140 V (3–4 V/cm) for 3–4 h.

4. After electrophoresis, rinse the gel with ddH2O for 30 min and transfer to hybond-N^+ membrane by capillary in 10× SSC overnight.

5. The next day, dissemble the capillary transfer unit, dry the membrane, and cross-link the membrane using auto cross-link twice in UV strataliker 1800.

6. Pre-hybridize the membrane in pre-hybridization buffer (*see* **Note 5**) at 68°C for 2–3 h in a hybridization tube with rotation using hybridization oven/shaker.

7. For hybridization probe preparation, dilute 25 ng DNA fragment containing hTER in 45 μl 1× TE. Denature the DNA sample by heating to 95–100°C for 10 min in boiling water. Snap cool the DNA in ice-water for 5 min after denature. Centrifuge briefly and add the denatured DNA to the reaction tube (RediPRIME II). Add 5 μl [^{32}P]dCTP, mix gently, and incubate at 37°C for 10 min. All radiochemicals should be handled according to radiation safety guidelines. Prepare Micro-Bio-Spin 6 chromatography columns by inverting the column several times to resuspend the gel and remove any bubbles. Snap off the tip, remove cap, and place the column in 2.0-ml collection tube to allow the excess buffer to drain by gravity. Discard the drained buffer and centrifuge for 2 min at $1000 \times g$ to remove the remaining packing buffer. Discard the drained buffer and place column in a new collection tube. Load the [^{32}P]dCTP-labeled probe and centrifuge for 4 min at $1000 \times g$. Keep the flowthrough and denature the probe at 95–100°C for 10 min in boiling water. Snap cool the probe in ice water for 5 min after denature.

8. Add the [^{32}P]dCTP-labeled hTER probe to membrane with pre-hybridization buffer and incubate at 68°C overnight.

9. To wash the membrane, drain the pre-hybridization buffer and wash the membrane twice with low-stringency wash buffer at room temperature. Then wash the membrane twice more in high-stringency wash buffer at 68°C. Dry the membrane and expose the membrane to X-ray film at various times to obtain sufficient exposure for detection of hTER (*see* **Fig. 2A** and **D**).

3.1.4. Quantification of Telomerase Activity Using Telomeric Repeat Amplification Protocol Assay

1. Telomeric repeat amplification protocol (TRAP) assay is performed using TRAPese telomerase detection kit. Briefly, 1×10^6 cells treated with lentivirus

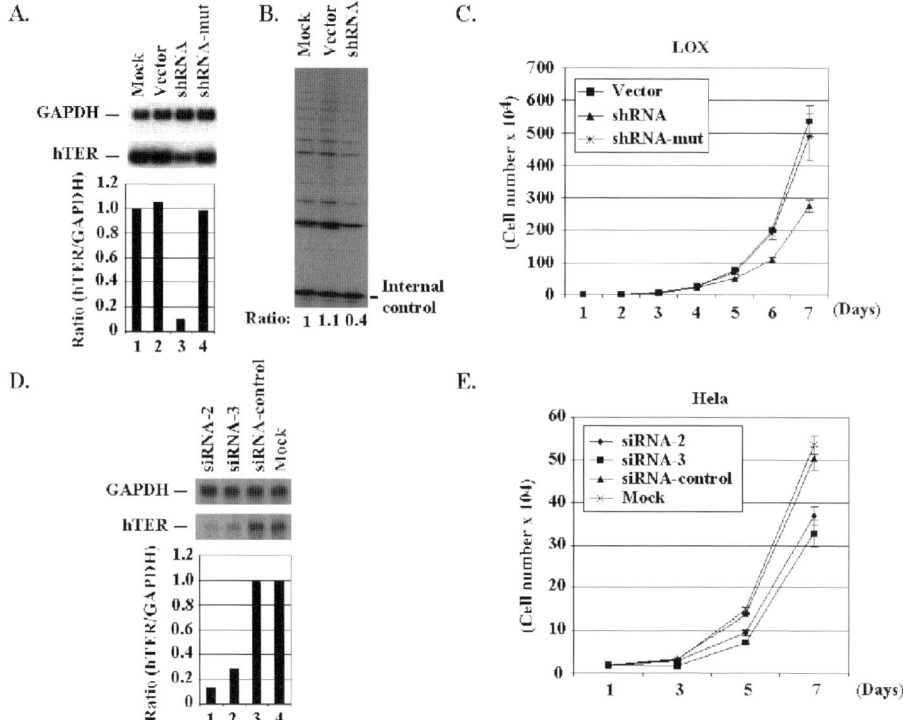

Fig. 2. Efficient knockdown of telomerase RNA and rapid cell growth inhibition induced by anti-hTER small-hairpin RNA (shRNA) and short-inference RNA (siRNA) in cancer cells. (**A**) Northern blotting analysis of LOX cells mock infected or infected with control lentivirus or lentivirus-expressing shRNA or shRNA-mut shows dramatic knockdown of endogenous hTER expression level by anti-hTER shRNA specifically. Endogenous glyceraldehyde-3-phosphate dehydrogenase (GAPDH) was used as the loading control. (**B**) Inhibition of telomerase activity in LOX cells by anti-hTER shRNA as shown by in vitro telomeric repeat amplification protocol (TRAP) assay. (**C**) Rapid cell growth inhibition in LOX cells infected with lentivirus-expressing anti-hTER shRNA but not cells infected with control lentivirus or lentivirus-expressing shRNA-mut targeting sequence unrelated to hTER. (**D**) Northern blotting analysis of Hela cells mock-transfected or transfected with control siRNA or anti-hTER siRNAs shows dramatic knockdown of endogenous hTER expression level by anti-hTER siRNAs specifically. Endogenous GAPDH was used as the loading control. (**E**) Rapid cell growth inhibition induced by synthetic anti-hTER siRNAs but not control siRNA in Hela cells.

or synthetic siRNA as described in **Subheading 3.1.2** are harvested by trypsin, washed once with 1× PBS, and pelleted into 1.5 ml Eppendorf tube with centrifugation.

2. Cells are resuspended in 200 μl 1× CHAPS lysis buffer in 1.5 ml Eppendorf tubes and incubated on ice for 30 min. The samples are centrifuged at $12,000 \times g$ for 20 min at 4°C. The supernatant is transferred to a new tube.

3. For endlabeling the TS primer, prepare the reaction mix as follows and incubate at 37°C for 20 min.

$\gamma-{}^{32}$P-ATP (3000 Ci/mmol, 10 mCi/ml)	2.5
TS primer	10
10× kinase buffer	2
T4 polynucleotide kinase (10 units/μl)	0.5
dH$_2$O	5
	20 μl

4. For assay setup, prepare the master mix as follows

10× TRAP reaction buffer	5.0
50× dNTP mix	1.0
^{32}P-TS primer	2.0
TRAP primer mix	1.0
Taq DNA polymerase (5 units/μl)	0.4
dH$_2$O	38.6
	48 μl

5. Aliquot 48 μl master mix into RNase-free PCR tube. Add 2 μl test extract or control into each tube (*see* **Note 6**).

6. For PCR amplification, place the tube in the thermocycler PTC-200 and incubate at 30°C for 30 min. After the 30-min incubation, perform two-step PCR at 94°C for 30 s and 59°C for 30 s for 25 cycles.

7. For casting of 10% sequencing gel, clean glass plate thoroughly and apply sigmacote to one plate to ensure release after electrophoresis. Assemble the plates with 0.5-mm spacer using EZ cast gasket. Mix 40 ml SequaGel concentrate, 50 ml SequaGel diluent, and 10 ml SequaGel buffer with 0.8 ml fresh prepared 10% APS and 40 μl TEMED. Pour the gel immediately and allow to sit at room temperature for no less than 1 h.

8. Pre-run the plate in 1× TBE in gel electrophoresis apparatus (Model S3S) for 15–20 min at 1500 V.

9. Add $50 \mu l$ 2× sample loading buffer into each PCR tube. Denature the sample for 2 min at 95°C and load $5 \mu l$ sample for each reaction. Run the gel at 1800 V for 3–4 h until the Xylene Cyanol FF dye is about 1/3 to the bottom.

10. Take the gel off with pre-cut 3 M paper and wrap with Saran wrap before exposing to phosphoimager screen overnight. Image data are scanned using phosphoimager scanner Storm 840.

11. Quantitation of TRAP activity can be done using ImageQuant software (*see* **Figs 2B** and **3E**).

3.1.5. Measurement of Cell Growth

1. At 24 h after cells are treated with lentivirus-expressing shRNA or synthetic siRNA as described in **Subheading 3.1.2**, the cells are harvested by trypsin and pellet in 15 ml sterile disposable conical tube.

2. Cells are resuspended in fresh DMEM. An aliquot of cells is mixed with 0.4% trypan blue stain solution at 1:1 ratio. The unstained cells were counted using a hemacytometer under the microscope to obtain the total number of viable cells. About 2×10^4 cells/well are reseeded in a 6-well plate. The medium is changed every other day to maintain active cell growth.

3. The cell number in each well is counted everyday or every other day and plotted as shown in **Fig. 2C** and **E** (*see* **Note 7**).

4. When cells in each well reach greater than 50% confluence, the cells are harvested by trypsin and reseeded into 10-cm cell culture dish to maintain active cell growth.

3.2. Telomerase RNA Knockdown by Hammerhead Ribozyme

Hammerhead ribozyme is one of the smallest catalytic RNAs that have been identified *(30)*. It is about 30 nucleotides in length and composed of three base-paired helixes (I–III). It catalyzes the site-specific cleavage of a phosphodiester bond in the RNA target through base pairing of target RNA with helixes I and III. The only sequence requirement of the target RNA is the presence of GUC cleavage triplet. To study the anti-tumor activity of telomerase inhibition by systematically delivered ribozyme, several ribozymes were designed to target different regions of mouse telomerase RNA that contained the essential GUC cleavage sequence at nucleotides 79, 100, 180, and 405 of mouse telomerase RNA. Ribozyme 180 was chosen for in vivo study for its efficient knockdown of mouse telomerase RNA expression and inhibition of telomerase activity in vitro using assays described in **Subheadings 3.1.2–3.1.4**. We will focus our description and discussion of procedures for analysis of cellular effects of mouse telomerase RNA knockdown by anti-mTER ribozyme in vivo in this section.

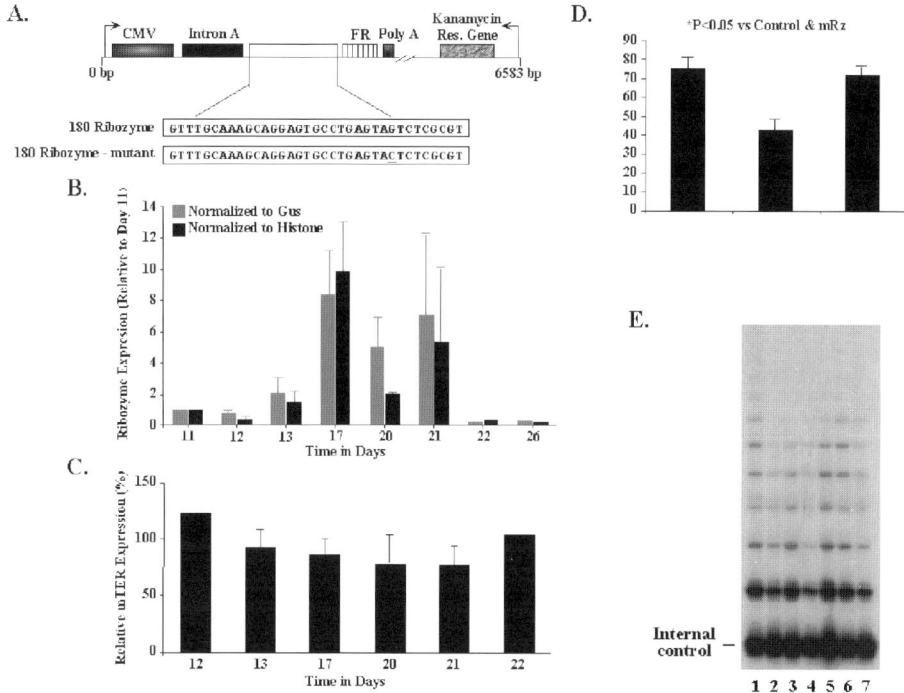

Fig. 3. Knockdown of murine telomerase RNA using systemically delivered ribozyme. (**A**) Schematic diagram of expression plasmid used to express anti-hTER ribozyme. FR, family repeats from Epstein–Barr virus (EBV). (**B**) Ribozyme expression levels in total cellular RNA from lungs of tumor-bearing mice by TaqMan analysis after cationic liposome : DNA complex-based delivery of plasmid encoding anti-hTER 180 ribozyme on experiment days 3 and 10 and killing mice on experiment days 11–26. (**C**) Telomerase RNA levels in total cellular RNA from lungs of tumor-bearing mice by TaqMan analysis on experiment days 11–26 under same condition as **B**. (**D**) Comparison of total lung tumor counts between mice treated with cationic liposome : DNA complexes containing vector control, anti-hTER 180 ribozyme, or anti-hTER 180 ribozyme-mutant on experiment day 27. *P value of the ribozyme-treated group when compared to the control or mRz-treated group. (**E**) Analysis of telomerase activity in metastatic lung tumors by telomeric repeat amplification protocol (TRAP) at 48 and 72 h after systematic treatment with cationic liposome : DNA complexes containing vector control or anti-hTER 180 ribozyme. Lane 1, 48-h post-vector therapy; lane 2, 48-h post-anti-hTER 180 ribozyme therapy; lane 3, 48-h post-vector therapy with higher lipid : DNA ratio; lane 4, 48-h post-anti-hTER 180 ribozyme therapy with same lipid : DNA ratio as lane 3; lanes 5 and 6, 72-h post-vector therapy; lane 7, 72-h post-anti-hTER 180 ribozyme therapy.

3.2.1. Engineer of Plasmids for Expression of Ribozyme

1. Plasmid 4613, CMV-Intron-FR-Kan, was constructed with the backbone plasmid VR1225, CMV-Intron-Luciferase-Kan *(31,32)*.
2. First, a 0.9-kb DNA fragment containing FR (the family of repeats, a major upstream enhancer of the Epstein–Barr virus) was inserted into the *Bam*H1 site.
3. Second, *luciferase* gene was excised with *Bam*H1 and *Eco*RV double digest. At each step, the plasmid was re-circularized by blunt-end ligation—staggered ends were blunted with Klenow.
4. Pair-wise oligomers (ribozyme 180 and ribozyme 180 mutant) were annealed at 85°C for 3 min, 65°C for 15 min, 37°C for 15 min, 25°C for 15 min, and 4°C for 15 min. The oligo pairs were then cloned into 4613 plasmid, linearized at *Pst*I/*Xba*I, by T4 DNA ligase (*see* **Fig. 3A**). The circular plasmids were transformed into subcloning efficiency DH5-α competent cells and grown in LB media with Kanamycin.
5. The DNA for in vivo use was isolated and purified with EndoFree Plasmid kit. The control sequence is without an insert.

3.2.2. Preparation of Cationic : Liposome Complexes

1. Liposome containing the cationic lipid DOTMA in a 1:1 molar ratio with cholesterol was prepared by drying down the lipid as a thin film in a 20×160 mm screw-capped glass tube using a rotary evaporator R-114 and then resuspending the dried lipid film in 5% w/v glucose by incubation in a water bath for 6 h at 54°C to produce multilamellar vesicle (MLV) liposome.
2. Small unilamellar vesicle (SUV) was produced from MLV by placing 1 ml MLV in a 16×160 mm crew-capped glass tube and sonicating the suspension in a cylindrical sonic bath for 15 min. The final 20 mM liposome solution is stored under argon at 4°C until use.
3. Prior to use, all solutions are allowed to normalize to room temperature. With 200-µl injection volume per mouse, 25 µg plasmid DNA is brought up to 100 µl with 5% dextrose. Similarly, 650 nmol 1,2-Dioleoyl-3-Trimethyl ammonium-Propane (DOTAP) is brought up to 100 µl with 5% dextrose, and both solutions are incubated at room temperature for 5 min. Then the DNA solution is added to the liposome solution, and the mix is allowed to incubate at room temperature for 20–30 min (*see* **Note 8**).

3.2.3. Systemic Delivery of Cationic : Liposome Complexes Containing Anti-mTER or Control Ribozyme to Tumor-Bearing Mice

1. Cell preparation: B16-F10 murine melanoma cells are grown in RPMI 1640, with 5% FBS at 37°C and 5% CO_2, in tissue-culture flasks up to 80% confluency (*see* **Note 9**).
2. On the day of injection, B16 cells are washed once with PBS and then detached with 0.05% trypsin (1 ml/75 cm^2), incubated at 37°C for 1–3 min. The trypsin is neutralized with 4 ml RPMI 1640 with 10% FBS. Cell concentration is measured with the help of

a hemocytometer, and the target cell-count concentration is achieved by appropriate amount of RPMI-only solution. About 25,000 B16-F10 cells are prepared in 200 ml volume, kept on ice, and injected immediately by tail vein injection.

3. The procedure for in vivo injection takes less than 30 s, during which mice are restrained (not anesthetized), with their tails immobilized on a platform. The 28.5-gauge needles are used to administer 200–400 μl solution into the lateral tail vein in one continuous push. The same route is utilized for both systemic inoculation of the animal with tumor cells and the administration of the therapeutic agents. Prior to needle insertion, the site of injection and the platform are sterilized with 100% ethanol.

3.2.4. Quantitative Analysis of Ribozyme and Mouse Telomerase RNA Expression After Systematic Delivery of Ribozyme

1. Total RNA was extracted from the lungs of tumor-bearing mice using TRIzole reagent. The tissue is added to lysis buffer (volume calculated according to weight of tissue or number of cells) and dispersed with Polytron PT 1200 homogenizer, at setting 4–5 for 30–45 s, on ice. RNA was then extracted according to manufacture protocol (for detail, *see* **Subheading 3.1.3.**). The final RNA pellet is resuspended in RNase-free water and its concentration is determined by spectrophotometer and its quality is confirmed using a Bioanalyser.

2. For quantitative real-time PCR (TaqMan assay), contaminating DNA is removed with DNase-I, and 250 ng DNase-I digested total cellular RNA was used to synthesize cDNA.

3. RNA was reverse transcribed into cDNA with iScript, 300 ng in a 20 ul volume, using a final reaction concentration of $1\times$ GeneAmp PCR buffer II, 7.5 mM $MgCl_2$ solution, 1 mM each dNTP, 5 μM random primers, 0.4 unit/μl RNaseOUT Recombinant RNase Inhibitor, and 0.6 unit/μl Superscript II. The reaction was incubated at 25°C for 10 min, 48°C for 50 min, and 70°C for 15 min in thermocycler.

4. For TaqMan analysis, performed on an AB Prism 7900 sequence detection system, 5 μl cDNA was amplified using a final concentration of 500 nM (each) forward and reverse primers, 200 nM probe, 200 mM (each) dNTPs, $1\times$ buffer A, and 0.025 unit/ul AmpliTaq Gold DNA polymerase. The amplification was conducted at 95°C for 12 min and for 45 cycles of 95°C for 15 s and 60°C for 1 min (*see* **Fig. 3B** and **C** and **Note 10**.

3.2.5. Determination of Anti-Tumor Efficacy

1. First, euthanasia is performed by administering 100% CO_2 for > 5 min, followed by dissecting the lung tissue by bilateral thoracotomy.

2. The lung tissue is then fixed by inserting a blunt needle into the trachea and perfusing the lung with 10% buffered formalin.

3. The tumor load is then assessed by carefully counting all surface tumors, with the aid of a light stereomicroscope (*see* **Fig. 3D**).

4. Notes

1. Water from Millipore ultra-pure water purification system with a resistance of 18.2 MΩ-cm is used for buffer preparation.
2. Combining diluted lipofectamine 2000 with diluted DNA or synthetic siRNA should be completed within 30 min to avoid drop of activity.
3. For lentivirus infection, using > 20 TU/cell will substantially increase non-specific cellular toxicity. The incubation of lentivirus with cells can be extended to 24 h instead of 8 h to increase transduction efficiency.
4. For RNA preparation and northern hybridization assay, RNase-free environment is required. Use RNase-free Eppendorf tubes and filter-tip and handle all samples with gloves. Keep extracted RNA samples at −80°C thereafter.
5. Do not microwave pre-hybridization buffer during preparation. Precipitation will occur when stored at room temperature. Heating up the solution in a 65°C water bath with stirring will dissolve the precipitates.
6. For accurate quantification of TRAP activity, serial dilution of the cell extract (1:10–1:50) may be necessary.
7. For cell-growth measurement, each time point should be repeated at least thrice and each experiment should be repeated at least twice.
8. The complex is injected within the next 3 h. For longer procedures, cationic : liposome complex (CLDC) preparation should be staggered. About 200 ul CLDC is injected in one slow and continuous push through tail vein, as described in **Subheading 3.2.3.**
9. Cells are thawed and allowed to propagate for at least 72 h prior to intravenous injection.
10. For the probe and primers of each gene, using Primer Express software v1.5 with 6-FAM fluorophore on the 5′ end and the Black Hole Quencher 1 (BHQ1) on the 3′ end, reactions were optimized to have > 90% PCR efficiency before the expression analyses. Also, reverse transcriptase linearity tests revealed linear reverse transcription with 250–500 ng total cellular RNA/reverse transcriptase reaction. For each sample, run in triplicate, expression levels of the gene of interest (the ribozyme or mTER) and control genes (*Gus* or *Histone*) were determined by quantitative real-time PCR, with parallel analysis of no reverse transcriptase controls to rule out DNA contamination. For each sample, the cycle threshold of the gene of interest (ribozyme or mTER) was normalized against the cycle thresholds of *histone* or *Gus* control genes, before determination of relative gene expression levels between samples.)

References

1. Karlseder, J., Broccoli, D., Dai, Y., Hardy, S., and de Lange, T. (1999) p53- and ATM-dependent apoptosis induced by telomeres lacking TRF2. *Science* 283, 1321–1325.
2. Smogorzewska, A., and de Lange, T. (2002) Different telomere damage signaling pathways in human and mouse cells. *EMBO J.* 21, 4338–4348.
3. Blackburn, E. H. (2000) Telomere states and cell fates. *Nature* 408, 53–56.

4. Watson, J. D. (1972) Origin of concatemeric T7 DNA. *Nat. New Biol.* 239, 197–201.

5. Levy, M. Z., Allsopp, R. C., Futcher, A. B., Greider, C. W., and Harley, C. B. (1992) Telomere end-replication problem and cell aging. *J. Mol. Biol.* 225, 951–960.

6. Wright, W. E., Tesmer, V. M., Huffman, K. E., Levene, S. D., and Shay, J. W. (1997) Normal human chromosomes have long G-rich telomeric overhangs at one end. *Genes Dev.* 11, 2801–2809.

7. Lee, H. W., Blasco, M. A., Gottlieb, G. J., Horner, J. W., 2nd, Greider, C. W., and DePinho, R. A. (1998) Essential role of mouse telomerase in highly proliferative organs. *Nature* 392, 569–574.

8. Blasco, M. A., Lee, H. W., Hande, M. P., Samper, E., Lansdorp, P. M., DePinho, R. A., and Greider, C. W. (1997) Telomere shortening and tumor formation by mouse cells lacking telomerase RNA. *Cell* 91, 25–34.

9. Bodnar, A. G., Ouellette, M., Frolkis, M., Holt, S. E., Chiu, C. P., Morin, G. B., Harley, C. B., Shay, J. W., Lichtsteiner, S., and Wright, W. E. (1998) Extension of life-span by introduction of telomerase into normal human cells. *Science* 279, 349–352.

10. Bryan, T. M., Englezou, A., Dalla-Pozza, L., Dunham, M. A., and Reddel, R. R. (1997) Evidence for an alternative mechanism for maintaining telomere length in human tumors and tumor-derived cell lines. *Nat. Med.* 3, 1271–1274.

11. Greider, C. W., and Blackburn, E. H. (1985) Identification of a specific telomere terminal transferase activity in Tetrahymena extracts. *Cell* 43, 405–413.

12. Yu, G. L., Bradley, J. D., Attardi, L. D., and Blackburn, E. H. (1990) In vivo alteration of telomere sequences and senescence caused by mutated Tetrahymena telomerase RNAs. *Nature* 344, 126–132.

13. Lendvay, T. S., Morris, D. K., Sah, J., Balasubramanian, B., and Lundblad, V. (1996) Senescence mutants of Saccharomyces cerevisiae with a defect in telomere replication identify three additional EST genes. *Genetics* 144, 1399–1412.

14. Cohn, M., and Blackburn, E. H. (1995) Telomerase in yeast. *Science* 269, 396–400.

15. Singer, M. S., and Gottschling, D. E. (1994) TLC1: template RNA component of Saccharomyces cerevisiae telomerase. *Science* 266, 404–409.

16. Rudolph, K. L., Chang, S., Lee, H. W., Blasco, M., Gottlieb, G. J., Greider, C., and DePinho, R. A. (1999) Longevity, stress response, and cancer in aging telomerase-deficient mice. *Cell* 96, 701–712.

17. Vulliamy, T., Marrone, A., Goldman, F., Dearlove, A., Bessler, M., Mason, P. J., and Dokal, I. (2001) The RNA component of telomerase is mutated in autosomal dominant dyskeratosis congenita. *Nature* 413, 432–435.

18. Feng, J., Funk, W. D., Wang, S. S., Weinrich, S. L., Avilion, A. A., Chiu, C. P., Adams, R. R., Chang, E., Allsopp, R. C., Yu, J., and et al. (1995) The RNA component of human telomerase. *Science* 269, 1236–1241.

19. Nakamura, T. M., Morin, G. B., Chapman, K. B., Weinrich, S. L., Andrews, W. H., Lingner, J., Harley, C. B., and Cech, T. R. (1997) Telomerase catalytic subunit homologs from fission yeast and human. *Science* 277, 955–959.

20. Kim, N. W., Piatyszek, M. A., Prowse, K. R., Harley, C. B., West, M. D., Ho, P. L., Coviello, G. M., Wright, W. E., Weinrich, S. L., and Shay, J. W. (1994) Specific association of human telomerase activity with immortal cells and cancer. *Science* 266, 2011–2015.

21. Wright, W. E., Piatyszek, M. A., Rainey, W. E., Byrd, W., and Shay, J. W. (1996) Telomerase activity in human germline and embryonic tissues and cells. *Dev. Genet.* 18, 173–179.

22. Masutomi, K., Yu, E. Y., Khurts, S., Ben-Porath, I., Currier, J. L., Metz, G. B., Brooks, M. W., Kaneko, S., Murakami, S., DeCaprio, J. A., Weinberg, R. A., Stewart, S. A., and Hahn, W. C. (2003) Telomerase maintains telomere structure in normal human cells. *Cell* 114, 241–253.

23. Shay, J. W., and Bacchetti, S. (1997) A survey of telomerase activity in human cancer. *Eur. J. Cancer* 33, 787–791.

24. Li, S., Crothers, J., Haqq, C. M., Blackburn, E. H., Nosrati, M., Bagheri, S., Ginzinger, D., Debs, R. J., Kashani-Sabet, M., Rosenberg, J. E., Donjacour, A. A., Botchkina, I. L., Hom, Y. K., and Cunha, G. R. (2005) Cellular and gene expression responses involved in the rapid growth inhibition of human cancer cells by RNA interference-mediated depletion of telomerase RNA. *J. Biol. Chem.* 280, 23709–23717.

25. Li, S., Rosenberg, J. E., Donjacour, A. A., Botchkina, I. L., Hom, Y. K., Cunha, G. R., and Blackburn, E. H. (2004) Rapid inhibition of cancer cell growth induced by lentiviral delivery and expression of mutant-template telomerase RNA and anti-telomerase short-interfering RNA. *Cancer Res.* 64, 4833–4840.

26. Nosrati, M., Li, S., Bagheri, S., Ginzinger, D., Blackburn, E. H., Debs, R. J., and Kashani-Sabet, M. (2004) Antitumor activity of systemically delivered ribozymes targeting murine telomerase RNA. *Clin. Cancer Res.* 10, 4983–4990.

27. Hahn, W. C., Stewart, S. A., Brooks, M. W., York, S. G., Eaton, E., Kurachi, A., Beijersbergen, R. L., Knoll, J. H., Meyerson, M., and Weinberg, R. A. (1999) Inhibition of telomerase limits the growth of human cancer cells. *Nat. Med.* 5, 1164–1170.

28. Fire, A., Xu, S., Montgomery, M. K., Kostas, S. A., Driver, S. E., and Mello, C. C. (1998) Potent and specific genetic interference by double-stranded RNA in Caenorhabditis elegans. *Nature* 391, 806–811.

29. Filipowicz, W. (2005) RNAi: the nuts and bolts of the RISC machine. *Cell* 122, 17–20.

30. Doherty, E. A., and Doudna, J. A. (2000) Ribozyme structures and mechanisms. *Annu. Rev. Biochem.* 69, 597–615.

31. Tu, G., Kirchmaier, A. L., Liggitt, D., Liu, Y., Liu, S., Yu, W. H., Heath, T. D., Thor, A., and Debs, R. J. (2000) Non-replicating Epstein-Barr virus-based plasmids

extend gene expression and can improve gene therapy in vivo. *J. Biol. Chem.* 275, 30408–30416.

32. Kashani-Sabet, M., Liu, Y., Fong, S., Desprez, P. Y., Liu, S., Tu, G., Nosrati, M., Handumrongkul, C., Liggitt, D., Thor, A. D., and Debs, R. J. (2002) Identification of gene function and functional pathways by systemic plasmid-based ribozyme targeting in adult mice. *Proc. Natl. Acad. Sci. U. S. A.* 99, 3878–3883.

11

Evaluation of Tankyrase Inhibition in Whole Cells

Tomokazu Ohishi, Takashi Tsuruo, and Hiroyuki Seimiya

Summary

The telomeric poly(ADP-ribose) polymerase (PARP), tankyrase 1, modulates the impact of telomerase inhibition on human cancer cells. Thus, overexpression of tankyrase 1 in telomerase-positive cancer cells confers resistance to telomerase inhibitors, such as MST-312, whereas pharmacological inhibition of tankyrase 1 enhances telomere shortening by MST-312. These facts indicate that tankyrase 1 could be a target for telomere-directed molecular cancer therapy. Here, the authors describe a convenient method to monitor the telomeric function of tankyrase 1. This protocol takes much less time than the telomere Southern blot analysis and can be utilized as a rapid screening system for tankyrase 1 inhibitors that are effective in intact cells. For direct monitoring of tankyrase 1 PARP activity, a protocol for the in vitro enzyme assay is also described.

Key Words: Telomere; telomerase; tankyrase 1; PARP; TRF1; electroporation; immunofluorescence microscopy; drug screening.

1. Introduction

Telomere elongation by telomerase is regulated in cis by a protein-counting mechanism, in which longer telomeres have larger amounts of *t*elomeric *r*epeat-binding *f*actor 1 (TRF1), a negative regulator for telomere length *(1)*. The double-strand telomere DNA-bound TRF1, in conjunction with the downstream TIN2 (*T*RF1-*i*nteracting *n*uclear factor 2)/TPP1 (originally named as *T*INT1, *P*TOP, or *P*IP1)/POT1 (*p*rotection *o*f *t*elomeres 1) telomere binding protein complex, blocks access of telomerase to its substrate, the 3- overhang of telomeric DNA (*see* **Fig. 1**). Shorter telomeres, however, have fewer TRF1

From: *Methods in Molecular Biology, vol. 405: Telomerase Inhibition*
Edited by: L. G. Andrews and T. O. Tollefsbol © Humana Press Inc., Totowa, NJ

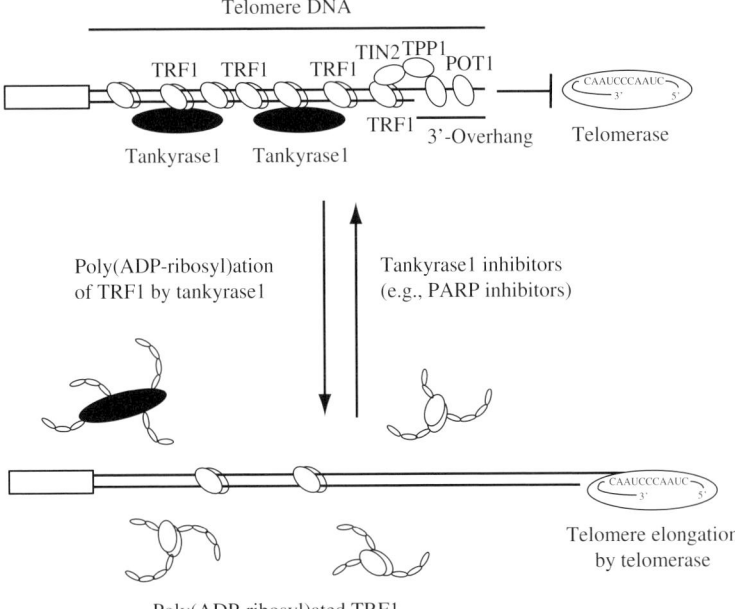

Fig. 1. Tankyrase 1 as a positive regulator of telomere elongation by telomerase. Human double-strand telomeric DNA is recognized by TRF1 dimers. TIN2 is a TRF1-binding protein that recognizes another protein, TPP1. TPP1 tethers POT1 to the single-strand overhang of telomere DNA. Thus, the TRF1/TIN2/TPP1/POT1 complex negatively regulates telomerase access to telomeres. In general, longer telomeres have more TRF1 and become less reactive substrates for telomerase. Tankyrase 1 poly(ADP-ribosyl)ates TRF1 and this post-translational modification leads to the dissociation of TRF1 from telomeres and its proteasomal degradation. As a result, enhanced access of telomerase increases the size of telomeres.

molecules, and thus telomerase has easier access. Consistent with this idea, telomere shortening per se gradually reduces the rate of telomere shortening by means of a telomerase inhibitor and attenuates the anti-proliferative impact of the drug on telomerase-positive cancer cells *(2)*.

Tankyrase 1 is a member of the poly(ADP-ribose) polymerase (PARP) family that catalyzes the addition of ADP-ribose polymers on acceptor proteins *(3)*. Tankyrase 1 poly(ADP-ribosyl)ates TRF1 and this post-translational modification releases TRF1 from telomere DNA *(4)* leading to proteasomal degradation of TRF1 *(5)*. Overexpression of tankyrase 1 in the nucleus of telomerase-positive cells decreases telomere binding of TRF1 *(6)*, increases the size of telomeres *(4)*, and confers resistance to telomerase inhibitors *(2)*. PARP inhibitors that can inhibit tankyrase 1 enhance telomere shortening by means of a telomerase inhibitor and thereby hasten the onset of cancer cell death *(2)*. These observations suggest tankyrase 1 as a potential target for telomere-directed molecular cancer therapeutics.

Measuring telomere length by conventional Southern blot and quantitative fluorescence in situ hybridization (FISH) analyses are the most straightforward approaches to monitoring the effects of test compounds on telomere dynamics. These techniques, however, require time to obtain meaningful results because detectable reduction in telomere length results from repetitive occurrence of the "end replication problem" of linear chromosomal DNA at each cell division. Here, the authors describe timesaving protocols for monitoring the telomeric function of tankyrase 1 (*see* **Fig. 2**). These techniques could be applied for screening of tankyrase 1 inhibitors in intact cells and for elucidation of intracellular regulatory pathways for tankyrase 1.

2. Materials

2.1. Cell Culture and Electroporation

1. Growth medium: Dulbecco's modified Eagle's medium (DMEM) (cat. no. 05919, Nissui, Tokyo, Japan) supplemented with 10% heat-inactivated bovine serum (HyClone, Ogden, UT) and 100 µg/ml kanamycin (Meiji Seika, Tokyo, Japan). Store at 4 °C and warm to 37°C before use. This growth medium has been confirmed to work with HeLa I.2.11 cells and other media could be used for other cell lines.
2. Phosphate-buffered saline (PBS) (cat. no. 05913, Nissui).
3. Trypsin–ethylenediamine tetra-acetic acid (EDTA) (0.05%) (cat. no. 25300-054, Gibco Invitrogen, Grand Island, NY).
4. Transformation buffer for electroporation; 21 mM HEPES, pH 7.05, 137 mM NaCl, 0.7 mM Na_2HPO_4, 5 mM KCl, 6 mM glucose. Filtrate with 0.45-µm filter and store at 4 °C.
5. Hemacytometer (Reichert, Buffalo, NY).
6. Plasmid DNA for tankyrase 1 expression vector, tagged with a FLAG epitope and a nuclear localization signal at the amino terminus (pcDNA3/FN-tankyrase 1) (*see* **Note 1**), is prepared using a QIAprep spin miniprep kit (cat. no. 27106, Qiagen, Hilden, Germany). DNA concentration should be higher than 0.5 mg/ml.
7. Gene Pulser with Capacitance Extender (Bio-Rad Laboratories, Hercules, CA).
8. Sterile cuvettes for electroporation (0.4-cm gap; cat. no. 72004, BioSmith, San Diego, CA, or equivalents).

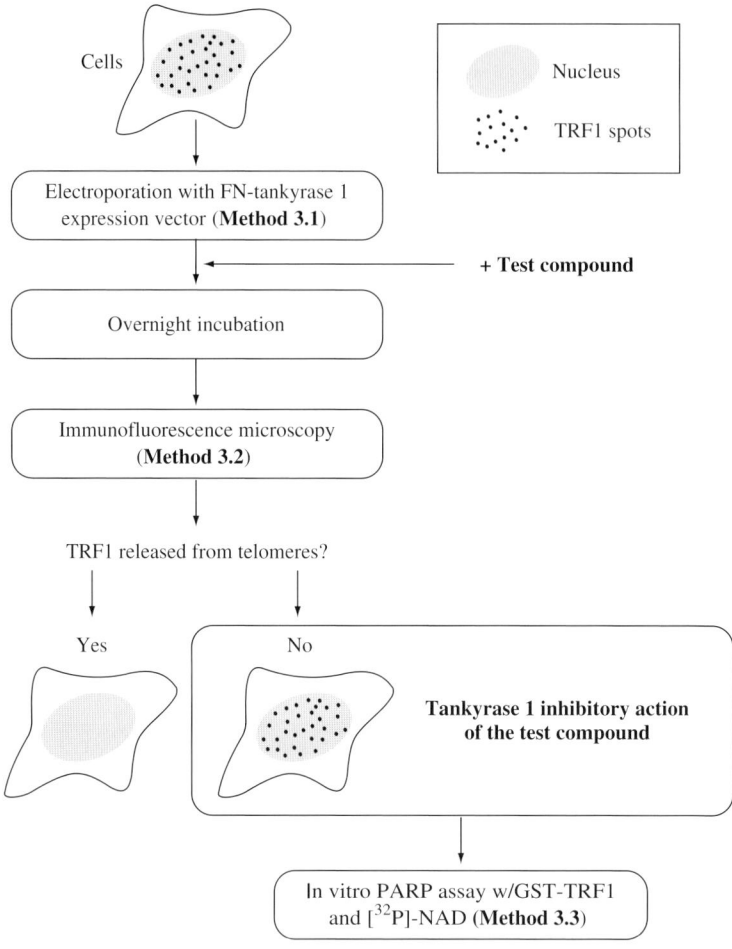

Fig. 2. Flow chart for the rapid monitoring of the telomeric function of tankyrase 1. Cells are transfected with an expression vector for tankyrase 1 (FN-tankyrase 1), which is tagged with a FLAG epitope and a nuclear localization signal for immunological detection and nuclear localization, respectively (**Method 3.1**). Electroporated cells are cultivated overnight in the presence of test compounds. Immunofluorescence staining of the cells with anti-FLAG and anti-TRF1 antibodies detects the loss of telomeric TRF1 signals in FN-tankyrase 1-overexpressing cells (**Method 3.2**). If the test compound has an ability to inhibit tankyrase 1 directly or indirectly, TRF1 will remain on the telomeres. To determine whether the compound directly inhibits the PARP activity of tankyrase 1, an in vitro enzyme assay can be performed (**Method 3.3**).

9. Circular glass cover slips, 12 mm in diameter (cat. no. 1912-002, Iwaki, Chiba, Japan); put into a glass dish with a lid, wrap in aluminum foil, and autoclave. It is not necessary to wash the cover slips before use. Handle with sterile, fine forceps.

10. 6-well flat-bottom culture plates (cat. no. 3810-006, Iwaki).

11. 3-aminobenzamide (3AB) (cat. no. A0788, Sigma, St. Louis, MO) and 3-aminobenzoic acid (cat. no. A9878, Sigma) are dissolved in dimethylsulfoxide at 3 M, stored at −20°C, and then add to tissue-culture plates as required.

2.2. Indirect Immunofluorescence Staining

1. 2% paraformaldehyde/PBS(−): dissolve 10 g paraformaldehyde (cat. no. 160-00515, Wako, Osaka, Japan) in 100 ml H_2O (*see* **Note 2**). Add 2 ml 0.2 N NaOH, heat to 60°C in fume hood, and stir until completely dissolved. Add 50 ml 10× PBS(−) and 350 ml H_2O. Store in 50 ml aliquots at −20°C.

2. 0.5% Nonidet P-40 (NP-40)/PBS(−): make 10% (v/v) NP-40 (cat. no. 23640-94, Nacalai Tesque, Kyoto, Japan) with H_2O. The solution is stored at room temperature. 0.5% NP-40/PBS(−) solution: dilute the 10% stock solution 1:20 with PBS(−).

3. 0.2% sodium azide solution: dissolve 0.2 g sodium azide (cat. no. 31233-42, Nacalai Tesque) in 100 ml H_2O. Stored at room temperature.

4. 1% bovine serum albumin (BSA)/PBS(−): dissolve 10 g BSA (Fraction V, cat. no. 735-078, Boehringer Manheim, Manheim, Germany) in 100 ml PBS(−) containing 0.02% sodium azide. The 10× stock solution is stored at 4°C. 1% BSA/PBS(−) solution: dilute the 10× stock solution 1:10 with PBS(−).

5. Primary antibodies: mouse anti-FLAG® M2 monoclonal antibody (cat. no. F3165, Sigma) and rabbit anti-TRF1 polyclonal antibody, affinity-purified, 5747 (*7*).

6. Secondary antibodies: anti-mouse immunoglobulin (Ig), fluorescein-linked whole antibody (from sheep) and anti-rabbit Ig, Texas red-linked whole antibody (from donkey) (N1031 and N2034, respectively, Amersham Biosciences, Piscataway, NJ).

7. VECTASHIELD® mounting medium with 1.5 μg/ml DAPI (4′,6-diamino-2-phenylindole cat. no. H-1200, Vector Laboratories, Burlingame, CA).

8. Microscope slide glasses, precleaned (cat. no. 12-552, Fisher Scientific, Pittsburgh, PA, or equivalents).

9. Clear nail polish.

2.3. In Vitro PARP Assay

1. HTC75 cells overexpressing FN-tankyrase 1: fibrosarcoma HTC75 cells were established in the Titia de Lange Laboratory, The Rockefeller University (*8*). These cells were infected with a retrovirus expressing FN-tankyrase 1, and the stable transformants were selected by exposure to 2 μg/ml puromycin for a week and maintained with the puromycin-containing growth medium (*7*).

2. Buffer C/NP-40: 20 mM HEPES, pH 7.9, 0.42 M KCl, 25% glycerol, 0.1 mM EDTA, 5 mM $MgCl_2$. Store at 4°C. Just before use, add 1/1000 volume of 1 M dithiothreitol (DTT; cat. no. 592-03951, Wako; the 1 M solution is

stored at −20 °C), 1/40 volume of the protease inhibitor cocktail (cat. no. P8340, Sigma), and 1/50 volume of 10% NP-40 solution.

3. PARP buffer: 50 mM Tris–HCl, pH 8.0, 4 mM $MgCl_2$. Store at 4 °C. Just before use, add 1/5000 volume of 1 M DTT.

4. 2× PARP buffer: 100 mM Tris–HCl, pH 8.0, 8 mM $MgCl_2$. Store at 4°C. Just before use, add 1/2500 volume of 1 M DTT.

5. EZview™ Red ANTI-FLAG® M2 Affinity Gel (cat. no. F-2426, Sigma).

6. GST-TRF1 and GST (negative control) recombinant proteins are produced in *Escherichia coli* BL21-CodonPlus strain (Stratagene, La Jolla, CA) and purified with Glutathione-Sepharose 4B (Amersham Biosciences) *(9)*. More detailed protocols are available upon request. In brief, the bead-bound proteins are eluted by 50 mM Tris–HCl, pH 8.0, 10 mM reduced glutathione and dialyzed against PBS(−) twice at 4 °C. Protein concentrations are determined using Bio-Rad protein assay reagent.

7. [^{32}P]-nicotinamide adenine dinucleotide (NAD) solution: to the bottle of NAD (29.6 TBq/mmol, 50 µl) (cat. no. NEG-023X, PerkinElmer, Wellesley, MA), 3.3 µl of 1 mM NAD (non-radioactive) is added. Resulting solution is diluted 1:4 with distilled water and used as the [^{32}P]-NAD solution.

8. Sodium dodecylsulfate (SDS) sample loading buffer: 62.5 mM Tris–HCl, pH 6.8, 2% SDS, 10% glycerol, 5% 2-mercaptoethanol, 0.001% bromophenol blue.

9. SDS–polyacrylamide gel electrophoresis (PAGE) buffer: for 10× stock buffer, dissolve 30.3 g Tris, 144.1 g glycine, 10 g SDS in H_2O and adjust the final volume to 1 l. This buffer is used for PAGE in 1:10 dilution with H_2O.

10. PAG Mini "DAIICHI" gels, 4–20% gradient (cat. no. 301506, Daiichi Pure Chemicals, Tokyo, Japan) and the electrophoresis apparatus (cat. no. DPE-1020, Daiichi Pure Chemicals).

11. Full-range rainbow molecular weight markers (cat. no. RPN800, Amersham Biosciences).

12. CBB staining solution: 0.025% Coomassie brilliant blue, 45% methanol, 10% acetic acid.

13. CBB destaining solution: 45% methanol, 10% acetic acid.

14. Gel dryer (Bio-Rad) with an aspirator.

15. BAS imaging plate IIIs2040 (Fuji Photo Film, Kanagawa, Japan).

16. BAS 2500 imaging analyzer (Fuji Photo Film).

3. Methods

Telomeric function of tankyrase 1 can be monitored using the transient overexpression system in intact cells. When exogenous tankyrase 1 is overexpressed in the nucleus, TRF1 dissociates from telomeres in a tankyrase 1 PARP activity-dependent manner *(4,10)*. Using indirect immunofluorescence microscopy, this phenomenon can be visualized as the disappearance of the TRF1 foci colocalized with telomeres (*see* **Fig. 3A**). If the cells were exposed to tankyrase 1 inhibitory compounds (e.g., PARP inhibitors), TRF1 would not be poly(ADP-ribosyl)ated. Thus, in that case, the telomeric spots of TRF1 would be detected even in the presence of exogenous tankyrase 1. Several PARP inhibitors, which block poly(ADP-ribosyl)ation of TRF1 but do not affect tankyrase 1/TRF1 interaction, enhance colocalization of these two proteins at the telomeres *(2)*. In this assay, effects of other PARP members, such as PARP-1 and PARP-2, would be minimal, if any, as the cells express much higher amounts of exogenous tankyrase 1 and at least PARP-1 does not poly(ADP-ribosyl)ate TRF1 *(10)*. To determine whether the compound directly inhibits PARP activity of tankyrase 1, an in vitro enzyme assay designed to monitor poly(ADP-ribosyl)ation of a recombinant TRF1 protein is recommended (*see* **Fig. 3B**) *(7)*.

3.1. Electroporation

1. When HeLa I.2.11 cells approach confluence (70–80%), they are passaged with trypsin–EDTA to provide new maintenance cultures on 10-cm culture dishes (*see* **Note 3**). Approximately one 15-cm dish is required for each electroporation sample.
2. Using fine-edged forceps, sterile glass cover slips are placed into the bottom of each well of 6-well plates (approximately five cover slips per well).
3. Cells on 15-cm culture dishes are washed with 10 ml PBS(–) and trypsinized for 5 min at room temperature. Detached cells are resuspended in 10 ml growth medium with gentle pipetting. A small aliquot of the cell suspension is transferred onto a hemacytometer for cell counting.
3. Cell number is immediately counted through a microscope. The volume necessary to give 1×10^7 cells (for each electroporation sample) is calculated and transferred to a 15-ml centrifuge tube.
4. The tube containing the cells is centrifuged at $210 \times g$ for 3 min at 4 °C.
5. The plasmid DNA for FN-tankyrase 1 (15 µg) is placed at the bottom of a 0.4-cm electroporation cuvette.
6. After centrifugation of the tube, the supernatant is removed by aspiration. The cell pellet is resuspended in the transformation buffer ($1 \, ml/1 \times 10^7$ cells) with gentle pipetting.
7. Using a P-1000 micropipetter, 0.8 ml suspension is transferred into the DNA-containing cuvette, taking care to avoid making bubbles. The cell suspension is mixed with DNA in the cuvette by gentle pipetting.
8. Cells are electroporated at 280 V, 960 µF with Gene Pulser. The time range (τ) is usually between 11 and 14 ms.
9. The white coat at the top of the suspension (cell debris) is removed with a sterile pipette tip (*see* **Note 4**). Then, the rest of the cells in the cuvette are mixed directly but gently by pipetting.

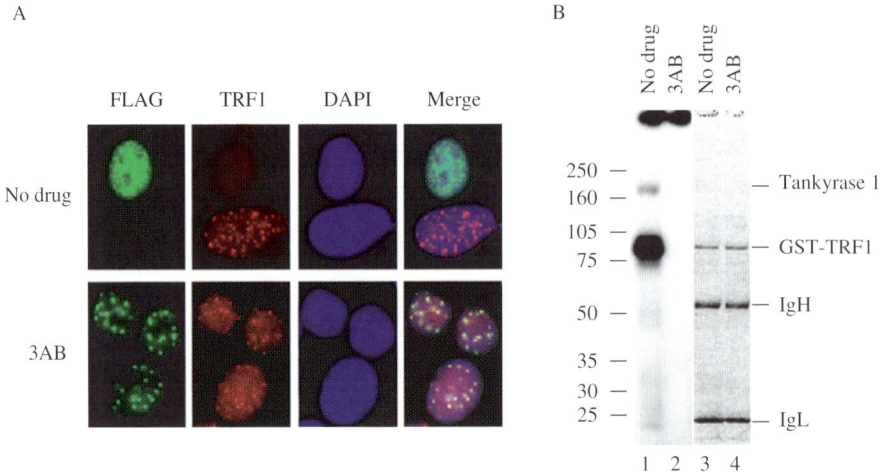

Fig. 3. Representative results of the tankyrase 1 assays. (**A**) Immunofluorescence staining of tankyrase 1 and TRF1. HeLa I.2.11 cells were transfected with FN-tankyrase 1 and fixed with paraformaldehyde. Exogenous FN-tankyrase 1 and endogenous TRF1 were detected with anti-FLAG and anti-TRF1 antibodies, respectively. DNA was counterstained with DAPI. Please note that FN-tankyrase 1-expressing cells lost the dot-like signal of TRF1 (*upper panels*). In the presence of a PARP inhibitor, 3AB (3 mM), FN-tankyrase 1 failed to dissociate TRF1 from telomeres (*lower panels*). (**B**) In vitro PARP assay of tankyrase 1. FN-tankyrase 1 was immunoprecipitated from cellular lysate and subjected to the PARP assay with a recombinant GST-TRF1 as an acceptor of ^{32}P-ADP-ribose. The reaction was subjected to SDS–PAGE, and specific signals were detected by autoradiography (*lanes* 1 and 2). Coomassie stain of the same gel is also shown (*lanes* 3 and 4). Poly(ADP-ribosyl)ated GST-TRF1 was detected in control but not in 3AB-containing reaction (1 mM). Auto-poly(ADP-ribosyl)ation of FN-tankyrase 1 was also detected in the control reaction. Molecular mass markers (kDa) are indicated at the left.

10. The cells are put in culture plates filled with glass cover slips (120 μl/well for 6-well plates). Then, the lid of the plate is closed, and the cells sit for 10 min at room temperature (*see* **Note 5**).

11. Growth medium is added to each well (4 ml/well for 6-well plates). The cells are mixed well by gentle pipetting. If needed, test compounds at appropriate concentrations are added to assigned wells. As a positive control for PARP inhibition, 3AB is added at the final concentration of 3 mM. As a negative control, 3-aminobenzoic acid (an analog of 3AB with no PARP inhibitory activity) at 3 mM can be used. The cells are mixed well by gentle pipetting.

12. The cells are cultured overnight in a CO_2 incubator at 37 °C. Then, the cell-attached cover slips are ready for indirect immunofluorescence staining.

3.2. Indirect Immunofluorescence Staining

1. The growth medium is removed from the culture plate containing the cell-attached cover slips. These cover slips are washed with PBS(–) (4 ml/well) once.

2. For fixation, 2% paraformaldehyde/PBS(–) (4 ml/well) is added to each well of the plate, and it is incubated at room temperature for 10 min. Then, each well is washed with PBS(–) for 2 min three times.

3. For permeabilization, 0.5% NP-40/PBS(–) (4 ml/well) is added to the wells, and the plate is incubated at room temperature for 10 min. Then, each well is washed with PBS(–) for 2 min three times. These cover slips can be stored in the presence of 0.02% NaN_3/PBS(–) at 4 °C for up to a month.

4. Using forceps, the cover slips to be stained are transferred (cell side up) to a chamber lined with parafilm, under which a piece of Whatman 3-MM chromatography paper wet with water has been placed.

5. To block non-specific background signals, 80 µl 1% BSA/PBS(–) is dropped onto each cover slip. Then, the lid of the chamber is closed, and the cover slips are incubated for 15 min at room temperature.

6. During the incubation, the primary antibody solution at working concentrations is prepared in a microtube: 2 µg/ml mouse anti-FLAG M2 monoclonal antibody and 1 µg/ml rabbit anti-TRF1 polyclonal antibody (5747) in 1% BSA/PBS(–) (*see* **Note 6**).

7. Using a vacuum pump (Bio Craft, BC-651, Tokyo, Japan, or an equivalent), the BSA/PBS(–) blocking solution is aspirated from each cover slip. Immediately after aspiration, the surface of the cover slip is filled with 80 µl 1% BSA/PBS(–) (*see* **Note 7**). The cover slips are incubated for 2 min at room temperature.

8. The BSA/PBS(–) solution is aspirated. The surface of the cover slip is filled with 80 µl 1% BSA/PBS(–) containing the primary antibodies. The lid of the chamber is closed, and the cover slips are incubated for 2 h at room temperature or overnight at 4°C (*see* **Note 8**).

9. During the incubation but just before using, a secondary antibody solution at working concentrations is prepared in a microtube: dilute both fluorescein-linked anti-mouse Ig and Texas Red-linked anti-rabbit Ig to 1:25 by 1% BSA/PBS(–). This antibody solution is kept in the dark until needed (e.g., wrapped in aluminum foil).

10. The primary antibody solution is aspirated from each cover slip. If the antibody solution is to be re-used in a future experiment, it is harvested into a microtube using a P-200 micropipetter. Each cover slip is washed with 1% BSA/PBS(–) for 2 min five times.

11. After the fifth BSA/PBS(–) washing solution is aspirated, the surface of the cover slip is filled with 80 µl secondary antibodies. The lid of the chamber is closed, and the cover slips are incubated in the dark for 45 min at room temperature.

12. During the incubation, the embedding medium, VECTASHIELD® mounting medium with DAPI, is removed from the refrigerator and settled at room

temperature (*see* **Note 9**). Meanwhile, appropriate numbers of glass slides are prepared and marked with experimental information (e.g., date of experiment, names of cells, treated drugs, and primary antibodies). Small drops of the embedding medium are put on the surface of the slide (*see* **Note 10**). Maximally, five cover slips can be put onto a single-standard glass slide.

13. The secondary antibody solution is aspirated from each cover slip. The cover slips are washed with 1% BSA/PBS(–) for 2 min five times.

14. After the fifth wash, the cover slips (cell side down) are placed onto the drops of the embedding medium on the glass slides. Excess amounts of the medium are removed by vacuum aspiration and/or blotted with soft paper towel (do not press the surface of cover slips firmly). The edges of the cover slips are sealed with clear nail polish, which is then let dry for 5–10 min at room temperature in the dark. The samples are subjected to fluorescence microscopy or stored at −20 °C (*see* **Note 11**).

3.3. In Vitro PARP Assay

1. HTC75 cells constitutively overexpressing FN-tankyrase 1 *(7)* are passaged like the HeLa I.2.11 cells. Approximately one 15-cm dish is needed for eight PARP reactions. Cells are trypsinized, resuspended in growth medium, harvested into a 15-ml tube, and centrifuged to collect the cell pellet. As a negative control, the parental HTC75 cells may be processed in the same way.

2. After the growth medium is removed by aspiration, the cell pellet is resuspended in ice-cold PBS(–) and transferred into a 1.5-ml microtube.

3. The microtube is centrifuged at $1500 \times g$ for 2 min at 4 °C. The supernatant is removed, and the cell pellet is resuspended in 1 ml ice-cold buffer C/NP-40 with gentle pipetting.

4. After placing on ice for 30 min, the microtube is centrifuged at $9000 \times g$ for 10 min at 4 °C. The supernatant (900 μl) is collected into a new microtube as a whole-cell extract. This extract can be stored at −80 °C.

5. For four PARP reactions, 450 μl cell extract is transferred into another microtube. Then, the microtube is supplemented with 30 μl (50% slurry) EZview™ Red ANTI-FLAG® M2 Affinity Gel, which has been prewashed once with ice-cold PBS(–) and twice with ice-cold buffer C/NP-40 and resuspended in the ice-cold buffer C/NP-40. The microtube is slowly rotated for 1 h at 4°C to form the FN-tankyrase 1/anti-FLAG immunocomplex (*see* **Note 12**).

6. The microtube is centrifuged at $13,000 \times g$ for several seconds at 4°C to precipitate the immunocomplex. The supernatant is removed by aspiration, and the beads are resuspended once in 1 ml ice-cold buffer C/NP-40 and twice in 1 ml ice-cold PARP buffer without DTT.

7. The immunocomplex is resuspended in 0.8 ml ice-cold PARP buffer containing 0.2 mM DTT. Then, the suspension is divided into four microtubes (i.e.,

200 μl/microtube). The microtubes are briefly centrifuged, and the supernatant is removed. Finally, the pellet is resuspended in 14 μl PARP buffer containing 0.2 mM DTT and 0.5 μg GST-TRF1 or GST (negative control) protein on ice (*see* **Note 13**). [Option] Before adding the [^{32}P]-NAD solution, if necessary, 1 μl 16 mM 3AB (final concentration 1 mM) or an appropriate amount of test compound is added to the reaction.

8. The PARP reaction is started by adding 2 μl [^{32}P]-NAD solution to the microtube (*see* **Note 14**). The reaction is performed at 25°C for 45–60 min, occasionally mixing the microtube by tapping.

9. SDS sample loading buffer at 5× concentration is added to the reaction (4 μl/tube). The cap of the microtube is rigidly closed, and the microtube is boiled at 100°C for 5 min, briefly spun down, and subjected to conventional SDS–PAGE. The authors use the 4–20% gradient polyacrylamide mini gel system available from Daiichi Pure Chemicals (Tokyo, Japan).

10. After PAGE, the electric power supply is turned off. The running buffer is carefully discarded into a tank of radioactive solution, according to safety guidelines. The electrophoresis apparatus is disassembled. The gel is submerged by the Coomassie brilliant blue staining solution for 20 min with gentle rocking. Then, the gel is submerged by the destaining solution for 1 h, washed with H$_2$O for 10 min four times. After washing, if necessary, the gel can be left in H$_2$O overnight at room temperature.

11. The gel is placed on a three-piece stack of Whatman filter paper, and the surface of the gel is covered with plastic wrap. The gel is vacuum-dried with a gel dryer at 80°C for 1–2 h.

12. The dried gel is exposed to a BAS imaging plate for 5–20 h at room temperature. The radioactive signal is quantitated with a BAS 2500 imaging analyzer. Alternatively, the gel is subjected to autoradiography at −80°C with an intensifying screen.

4. Notes

1. Nuclear accumulation of exogenously overexpressed tankyrase 1 requires an artificial NLS (nuclear localization signal) *(11)*. As tankyrase 1 is present not only at the telomeres but also at other loci outside the nucleus *(11–13)*, this modification allows us to enhance only the nuclear function of tankyrase 1 *(2,4–7,14)*. The FLAG epitope allows it to be detected specifically by anti-FLAG antibody.

2. Paraformaldehyde is highly toxic. Wear mask and gloves, and handle under a fume hood.

3. The authors' laboratory conventionally uses this cell line, which was established by the Titia de Lange laboratory, The Rockefeller University *(15)*. This cell line has very long telomeres and therefore exhibits very clear signals of telomeric TRF1

spots. Employing an immuno-FISH assay, we have confirmed that the TRF1 spots colocalize with telomere DNA (unpublished observation).

4. The cell debris tends to adhere and may give background signals in immunofluo-rescence staining unless properly removed at this point.

5. Do not disperse the cell suspension extensively in the well. It could hasten the liquid to evaporate.

6. If available amounts of antibody are limited, one can reduce the volume to less than 45 µl/cover slip and re-use the supernatant once more. In that case, the supernatant is collected into a microtube and stored at 4 °C. Multiple usage of the same antibody solution (i.e., more than three times) could lower the intensity of specific signals and give higher background signal. Typically, only the affinity-purified anti-TRF1 antibody gives fine telomeric spots of TRF1; un-purified, crude antiserum gives background stain in the nucleoplasm.

7. Do not allow the surface of the cover slip to dry. The authors usually remove the liquid with one hand, aspirating it with a needle tip, and supplie new liquid with the other hand, using a P-200 micropipetter, thus handling cover slips one by one.

8. In general, overnight incubation produces a little stronger signal than 2-h incubation. However, it may also give higher background signals.

9. Chilled embedding medium tends to produce small bubbles when cover slips are placed on it. These bubbles could disturb subsequent microscopic analysis.

10. As even one drop of the medium is too much for mounting a single cover slip, the medium may be transferred to a microtube (wrapped in aluminum foil) and put on slide glasses using a P-200 micropipetter.

11. Most of the cells exhibit multiple red spots of telomere-bound TRF1 in the nucleus (i.e., within the area of DAPI stain). Under these transfection conditions, approxi-mately 5–20% of the cells exhibit green signal of FN-tankyrase 1 in the nucleus. In FN-tankyrase 1-positive cells, TRF1 spots are not observed because TRF1 is released from telomeres *(4)* and degraded by the ubiquitin-proteasome pathway *(5)*. Some populations of the FN-tankyrase 1-positive cells may give aggregated forms of the exogenous tankyrase 1, which are presumably because of self-polymerization through its sterile alpha motif (SAM) domain *(16,17)*. Polymerized tankyrase 1 seems to have less ability to dissociate TRF1 from telomere DNA (Seimiya, H., unpublished data).

12. This immunoprecipitated tankyrase 1 can be utilized as a more convenient source of the enzyme as compared with relatively time-consuming baculoviral expression and purification systems.

13. Actually, this reaction buffer is made of 6 µl GST-TRF1 (or GST as negative control) and 8 µl 2× PARP buffer containing 0.4 mM DTT.

14. Before starting the experiments, make sure that there is no radioactive contam-ination on anything in or around the work area. For safety, transparent plastic should be used to shield against ^{32}P. When handling radioactive materials, follow legal and institutional regulations and instructions.

Acknowledgments

We thank Yukiko Muramatsu and Michiko Sakamoto for technical assistance and Susan Smith for scientific advice. This work was funded in part by a Grant-in-Aid for Scientific Research from the Ministry of Education, Culture, Sports, Science and Technology, Japan, and grants from the Kato Memorial Bioscience Foundation and the Vehicle Racing Commemorative Foundation, Japan.

References

1. Smogorzewska, A. and de Lange, T. (2004) Regulation of telomerase by telomeric proteins. *Annu Rev Biochem* **73**, 177–208.
2. Seimiya, H., Muramatsu, Y., Ohishi, T. and Tsuruo, T. (2005) Tankyrase 1 as a target for telomere-directed molecular cancer therapeutics. *Cancer Cell* **7**, 25–37.
3. Smith, S., Giriat, I., Schmitt, A. and de Lange, T. (1998) Tankyrase, a poly(ADP-ribose) polymerase at human telomeres. *Science* **282**, 1484–1487.
4. Smith, S. and de Lange, T. (2000) Tankyrase promotes telomere elongation in human cells. *Curr Biol* **10**, 1299–1302.
5. Chang, W., Dynek, J. N. and Smith, S. (2003) TRF1 is degraded by ubiquitin-mediated proteolysis after release from telomeres. *Genes Dev* **17**, 1328–1333.
6. Loayza, D. and de Lange, T. (2003) POT1 as a terminal transducer of TRF1 telomere length control. *Nature* **423**, 1013–1018.
7. Seimiya, H., Muramatsu, Y., Smith, S. and Tsuruo, T. (2004) Functional subdomain in the ankyrin domain of tankyrase 1 required for poly(ADP-ribosyl)ation of TRF1 and telomere elongation. *Mol Cell Biol* **24**, 1944–1955.
8. van Steensel, B. and de Lange, T. (1997) Control of telomere length by the human telomeric protein TRF1. *Nature* **385**, 740–743.
9. Seimiya, H. and Smith, S. (2002) The telomeric poly(ADP-ribose) polymerase, tankyrase 1, contains multiple binding sites for telomeric repeat binding factor 1 (TRF1) and a novel acceptor, 182-kDa tankyrase-binding protein (TAB182). *J Biol Chem* **277**, 14116–14126.
10. Cook, B. D., Dynek, J. N., Chang, W., Shostak, D. and Smith, S. (2002) A role for the related poly(ADP-ribose) polymerases tankyrase 1 and 2 at human telomeres. *Mol Cell Biol* **22**, 332–342.
11. Smith, S. and de Lange, T. (1999) Cell cycle dependent localization of the telomeric PARP, tankyrase, to nuclear pore complexes and centrosomes. *J Cell Sci* **112**, 3649–3656.
12. Chi, N.-W. and Lodish, H. F. (2000) Tankyrase is a Golgi-associated mitogen-activated protein kinase substrate that interacts with IRAP in GLUT4 vesicles. *J Biol Chem* **275**, 38437–38444.

13. Lyons, R. J., Deane, R., Lynch, D. K., Ye, Z.-S. J., Sanderson, G. M., Eyre, H. J., Sutherland, G. R. and Daly, R. J. (2001) Identification of a novel tankyrase through its interaction with the adaptor protein Grb14. *J Biol Chem* **276**, 17172–17180.

14. Ye, J. Z.-S. and de Lange, T. (2004) TIN2 is a tankyrase 1 PARP modulator in the TRF1 telomere length control complex. *Nat Genet* **36**, 618–623.

15. van Steensel, B., Smogorzewska, A. and de Lange, T. (1998) TRF2 protects human telomeres from end-to-end fusions. *Cell* **92**, 401–413.

16. De Rycker, M., Venkatesan, R. N., Wei, C. and Price, C. M. (2003) Vertebrate tankyrase domain structure and sterile-alpha motif (SAM) mediated multimerization. *Biochem J* **372**, 87–96.

17. De Rycker, M. and Price, C. M. (2004) Tankyrase polymerization is controlled by its sterile alpha motif and poly(ADP-ribose) polymerase domains. *Mol Cell Biol* **24**, 9802–9812.

12

Inhibition of Telomerase by Targeting MAP Kinase Signaling

Dakang Xu, He Li, and Jun-Ping Liu

Summary

Constitutive activation of the mitogen-activated protein (MAP) kinase signaling pathway by oncogenic stimulation is widespread in human cancers. With the recently demonstrated links between MAP kinase, histone phosphorylation, gene transcription factors, and *hTERT* gene promoter activity, abnormal MAP kinase activity is likely to be one of the essential forces that impact on *hTERT* gene transcription in transformed human cells. Several proteins have been implicated as playing important roles in MAP kinase signaling to *hTERT* gene, including *Ets* and activator protein-1 (*AP-1*). Inhibition of these signaling mechanisms may have a consequential effect on hTERT gene expression and telomerase activity. In this study, we brief the current progress and strategy in molecular targeting to the interface between MAP kinase and *hTERT* gene promoter in cancer.

Key Words: MAP kinase; telomerase; hTERT; signal transduction; cancer; inhibitor.

1. Introduction

In multicellular organism, hTERT expression is tightly controlled within the cell. Extracellular molecules communicate with nuclear transcription factors to start gene regulatory machinery through initiating intracellular signaling cascades. Occurring within cell and in between, signal transduction cascades are constantly regulated in temporal and spatial manners and various formats of interactions during the development, aging, and pathogenesis. Despite increasing knowledge on hTERT transcription regulation, still much less is

From: *Methods in Molecular Biology, vol. 405: Telomerase Inhibition*
Edited by: L. G. Andrews and T. O. Tollefsbol © Humana Press Inc., Totowa, NJ

known about which types of internal and external agents could interfere with hTERT gene regulation. Recent studies have shown that in epithelial cells, oncogenes such as Her2/Neu *(1)* or epidermal growth factor (EGF) *(2)* activate hTERT at gene transcription levels. In contrast, transforming growth factor-β (TGF-β) acts as a suppressor of hTERT gene expression in various cellular models *(3,4)*. In these regulations, however, the mitogen-activated protein (MAP) kinase signaling cascades are involved *(3,4)*. The MAP kinase cascades cooperate in transmitting extracellular signals to intracellular targets and initiate cellular processes such as proliferation, differentiation, stress response, transformation, and apoptosis. Each of these signaling cascades consists of three to six tiers of protein kinases that sequentially activate each other by phosphorylation. The similarity between the enzymes that comprise each tier in the various cascades makes them a part of super family of protein kinases (*see* **Fig. 1**).

The MAP kinase cascades are activated through membrane-bound Ras, with discontinuous relays of protein interactions and associated protein phosphorylation. The four or five tiers in each of the several MAP kinase cascades are

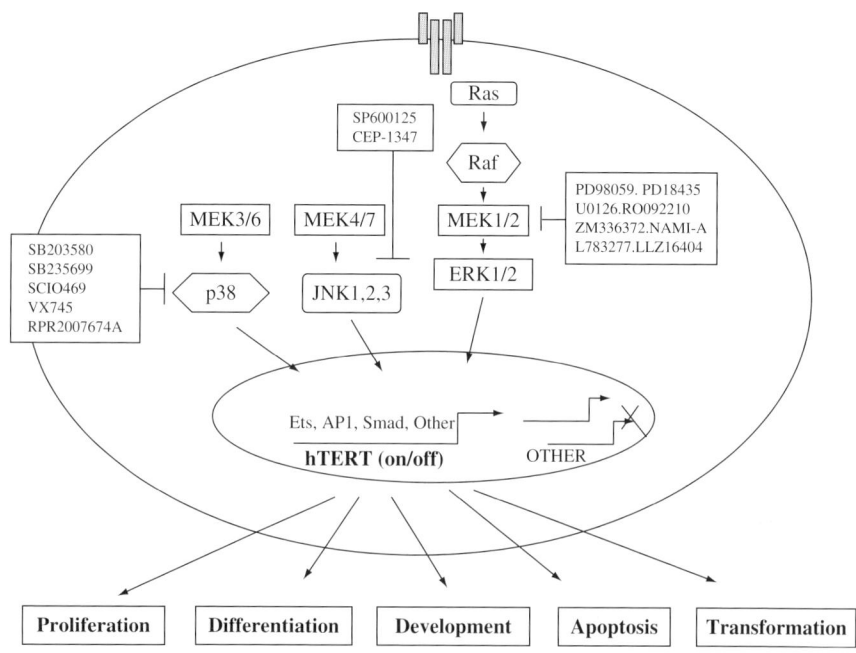

Fig. 1. Schematic presentation of mitogen-activated protein (MAP) kinase signaling pathways and events and their blockades by MAP kinase inhibitors in mammalian cells.

probably essential for signal amplification, specificity determination, and tight regulation of the transmitted signal. All the enzymes share common mechanisms of activation by protein phosphorylation, but at the MAP kinase level, the phosphorylation sites are threonine (Thr) and tyrosine (Tyr), arranged in a Thr-Xaa-Tyr motif that can be used to distinguish individual cascades. Multicellular organisms have four subfamilies of MAP kinases, namely extracellular signal-regulated kinase (ERK), c-Jun amino-terminal kinase (JNK), p38, and BMK (also known as ERK5) protein kinases, which control a vast array of physiological processes. The ERKs are involved in the control of both cell proliferation and cell differentiation. The JNKs also known as stress-activated protein kinase 1 (SAPK1) are critical regulators of apoptosis and gene transcription. The p38 protein kinases, also known as SAPK2–4, are activated by inflammatory cytokines and environmental stresses. In addition, some of the signaling cascades are cell type and condition specific and coordinated by interactions across different pathways in response to stress and oncogenic stimuli.

Activation of the Ras/Raf/MEK/MAP kinase pathway has been implicated in uncontrolled cell proliferation and tumor growth. Mutated, oncogenic forms of Ras are found in 50% of colon, 90% of pancreatic, and 30% of lung cancers. Recently, B-Raf mutations have been identified in more than 60% of malignant melanomas and from 40 to 70% of papillary thyroid cancers. MAP kinase kinase (MEK), a dual specificity kinase, is the key player in this pathway; it is downstream of both Ras and Raf and activates ERK1/2 through phosphorylation of key Tyr and Thr residues *(5)*. While the mechanisms by which the MAP kinase signaling cascades participate in tumorigenesis remain to be incompletely understood, especially on the roles of downstream molecules and specific effects of each pathway and their purposely particular combinations, it has been thought that MAP kinase signaling critically regulates one or several key events that are crucially required during oncogenic development including immortalization and transformation of proliferative cells. Given that telomere maintenance by telomerase is essentially required for stem cell renewal and immortalization of neoplastic cells, it is important to address any potential links between MAP kinase-mediated mitogenic signaling and telomerase-mediated tumor cell immortalization. Consistently, the autonomous activities in both mitogenic MAP kinase signaling and telomerase activity by mutations of oncogenes such as Ras, Raf, and Myc in most cancer cells mirror an intimate association between telomerase and MAP kinase families. This chapter briefly reviews the evidence of interrelationships between MAP kinase signaling and telomerase, with a focus on recent development and hypothesis, and in detail provides laboratory protocols in our studies of MAP kinase signaling and regulation of telomerase activity in cancer cells.

1.1. MAP Kinase Singling to hTERT Regulation in Cancer

As the key component of the telomerase complex, hTERT is tightly proliferation regulated and becomes activated in approximately 85% of most cancer types (*6*); however, underlying mechanisms are unclear. Recent studies have shown the presence of multiple mechanisms for the MAP kinase regulation of telomerase activity and identified functional Ets binding motif on the hTERT promoter through which Ras-ERK signaling may activate *hTERT* gene transcription and up-regulation of telomerase activity (*7*). Consistently, the oncogenic Ras has been shown to stimulate the hTERT promoter activity through the Ets transcription factor ER81 and MAP kinase (*1*). Characterized by an evolutionarily conserved Ets domain, Ets family of transcription factors plays important roles in cell development, differentiation, proliferation, and aging. Most of the Ets proteins are downstream nuclear targets of MAP kinase signaling, and the deregulation of Ets genes results in the malignant transformation of cells. Remarkably, three oncoproteins upstream of MAP kinase, HER2/Neu, Ras, and Raf, stimulate hTERT promoter activity through the Ets transcription factor ER81 and collaborate to enhance endogenous telomerase activity in telomerase-negative BJ foreskin fibroblasts (*1*). Mutating ER81 binding sites in the hTERT promoter renders *hTERT* gene promoter unresponsive to HER2/Neu, which can be reversed by expression of dominant-negative ER81 or inhibition of HER2/Neu signaling (*1*). Therefore, inhibition of the MAP kinase cascade may lead to diminished effect of the Ets transcription factor on *hTERT* gene transcription.

In addition, studies have shown that transactivation of the JNK induces telomerase activity (*8*). Arresting JNK by expression of the scaffold protein JNK-interacting protein-1/Islet-Brain1 (JIP-1/IB1) abrogates telomerase activity. The findings that JNK expression activates transcription of a reporter gene fused to the hTERT promoter sequence suggest that JNK regulates *hTERT* gene transcription (*8*). How JNK regulates *hTERT* gene transcription remains unknown. Although phosphorylation of c-Jun by JNK plays a critical role in transcriptional activation by activator protein-1 (AP-1), the putative AP-1 binding site at –1655 bp of the *hTERT* gene promoter is unlikely to be involved in the transcriptional activation of hTERT by JNK as AP-1 has recently been shown to be repressor of *hTERT* gene (see page 151). More recently, studies have found that MAP kinase mediates phosphorylation of histone H3, and phosphorylation at Ser10 connects cell proliferation with telomerase activation (*9*). In normal human T lymphocytes and fibroblasts, MAP kinase-induced phosphorylation of Ser10 in histone 3 on the hTERT promoter is critically associated with telomerase activity *(9)*. In addition, it is likely that MAP kinase signaling stimulates telomerase by multiple mechanisms including MAP kinase phosphorylation of histone 3, JNK, and Ets transcription factor ER81. It is possible that Ets transcription factors and histone H3 phosphorylation cooperate to activate the *hTERT* gene by both cis and trans mechanisms. Thus, MAP kinase signaling may be involved in regulating tumor cell immortalization by acting upon *hTERT* gene expression, telomerase activity, and telomere maintenance.

In the exquisite responses of MAP kinase signaling to the extracellular and intracellular environment, MAP kinase is also essentially involved in mediating cell differentiation that acts as a barer to oncogenesis *(7)*. It is thereby intriguing how MAP kinase signaling on the one hand stimulates cell proliferation and on the other hand promotes cell differentiation. Interestingly, recent studies provide evidence showing that MAP kinase can also transmit signal to repress *hTERT* gene transcription *(7,10,11)*. Thus, MAP kinase signaling dually up and down controls telomerase activity in a manner depending on detailed signaling composition. The transcription factor AP-1 involved in cellular proliferation, differentiation, and tumorigenesis binds to *hTERT* gene promoter directly repressing *hTERT* gene transcription in cancer cells. Combination of either c-Fos and c-Jun or c-Fos and JunD strongly represses hTERT promoter activity, specific effect in humans that is not observed on *mouse TERT* gene *(10)*. Consistent with both stimulatory and inhibitory roles played by the MAP kinase signaling, we observed that EGF stimulates *hTERT* gene transcription that is blocked by PD98059 (*see* **Figs. 2 and 3**) and that nerve growth factor (NGF) induces telomerase inhibition that is blocked by PD98059 *(7)*. In addition to AP-1 that might possibly mediate NGF-induced down-regulation of telomerase, it has been shown that Myc as a potent transcription factor of *hTERT* gene is down-regulated at gene transcription levels by NGF *(12)*. Thus, it is likely that depending on the extracellular environment, MAP kinase signaling is activated with specific downstream signaling configurations that impacts upon gene expressions for specifically different outcomes. MAP kinase may participate in suppression of hTERT and telomerase in regulating cell differentiation by multiple mechanisms involving AP-1 and Myc.

With the dual stimulatory and inhibitory signaling inputs into the hTERT locus of human genome, hTERT is regulated to control chromosomal integrity and stability by acting on the homeostasis of the chromosomal ends—telomeres. It is thus envisaged that different MAP kinase regulation of telomerase activity is associated with different status of telomere structures including

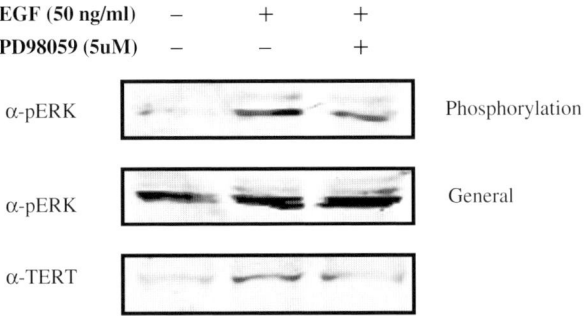

Fig. 2. Effect of the extracellular signal-regulated kinase (ERK) inhibitor PD98059 on activation of ERK and gene expression of hTERT stimulated by epidermal growth factor (EGF) stimulation in human mammary carcinoma MCF-7 cells. Phosphorylated ERK (pERK) was determined by western blotting with anti-ERK phospho-specific antibodies. Used was 25 ug protein extracts from resting or stimulated cells by EGF (50 ng/ml) plus or minus PD98059 (5 uM) as indicated in each condition. Lower panels show the blotted membranes probed with ERK antibody for total ERK protein and hTERT antibody for hTERT expression.

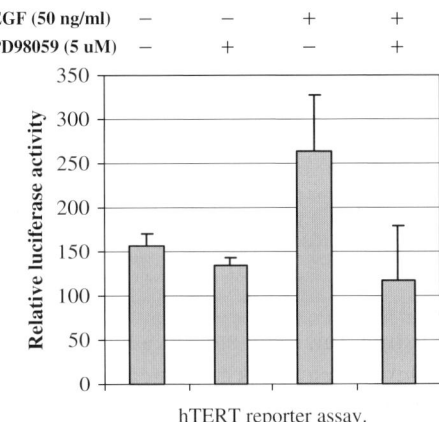

Fig. 3. Regulation of hTERT promoter by mitogen-activated protein (MAP) kinase signaling. MCF-7 cells were transfected with wild-type hTERT promoter luciferase reporter gene plasmids and then treated with or without epidermal growth factor (EGF) (50 ng/ml) with or without PD98059 (5 uM) as indicated for 24 h followed by luciferase activity assayed.

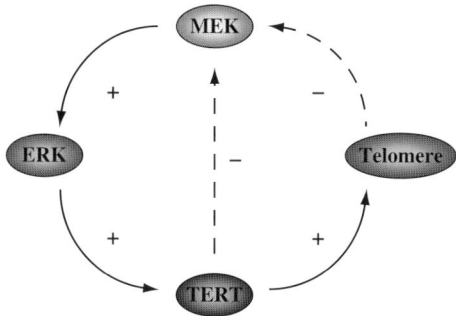

Fig. 4. Model of bi-directional regulations between mitogen-activated protein (MAP) kinase signaling and telomerase maintenance of telomeres. MEK activation of extracellular signal-regulated kinase (ERK) leads to telomerase activation through histone 3 phosphorylation *(9)* and ER81 binding and activation of *hTERT* gene transcription *(1)*. TERT instigates telomerase activity and telomere maintenance, but telomerase activity inhibits MEK activity *(12)*, implementing a feedback control loop. The mechanism of telomerase inhibition of MEK could be direct or indirect through telomeres.

the length of telomeric DNA that has been compellingly shown to regulate cell-proliferative lifespan. However, it was intriguingly surprising to note another twist of research findings from a recent study in that expression of hTERT plays a remarkable role in controlling MAP kinase signaling *(13)*, literally providing a negative feedback loop in switching on and off MAP kinase signaling–telomerase axis (*see* **Fig. 4**). In cultured bovine lens epithelial cells (BLECs), expression of hTERT suppresses ERK1/2 activity, and this inhibition is mediated by hTERT-induced inhibition of MEK1/2 *(13)*. Moreover, by suppression of MEK-MAP kinase singling, hTERT inhibits BLEC differentiation *(13)*, which is in contrast to that in PC-12 cells where hTERT is observed to play no role in mediating NGF-induced neural differentiation *(7)*. Although the mechanisms of hTERT feedback regulation of MAP kinase signaling requires investigation, these studies suggest that targeting MAP kinase signaling at different levels and sites is critical in modifying particular intracellular events and cellular activity.

1.2. Targeting the Interface Between MAP Kinase Singling and hTERT Gene

The Ras-Raf-MEK-ERK intracellular signaling pathway can be activated in response to a variety of extracellular stimuli. Growth factor binding to extracellular receptors results in activation of Ras, which in turn interacts with and activates Raf, leading to the phosphorylation of the dual specificity kinase MEK on two distinct serine residues. MEK possesses a number of unique biochemical and biological features that make it an attractive target from anti-cancer drug-development perspectives. The identification and subsequent testing of highly selective small molecule inhibitors of MEK have served to possibly realize the long-held belief of targeting the MEK/ERK module to modulate a number of cellular events that are critical to tumor cell proliferation and survival. Pfizer has advanced the first MEK-targeted clinical drug candidate into clinical trials with the entry of CI-1040. Evaluation of sufficiently potent and selective MEK inhibitors in well-designed clinical trials will be critical for ultimate validation of MEK as a molecular-based anticancer drug target. However, an important issue regarding the different MAP kinase inhibitor approaches on trial currently is that none of them is truly selective for cancer cells. The caveat is that normal Ras and its signaling pathways are also vital for normal cell physiology. Hence, it is anticipated that cancer cells are more dependent on Ras signaling, such that partial inhibition by drug treatment would be sufficient to impede oncogenesis. There is some suspicion that the function of mutated Ras does not simply represent excessive activity of the normal Ras; if so, this may provide another window to the selectivity of future anti-cancer drug by screen MAP kinase inhibitor from hTERT-positive cancer cells.

Studies have shown that hTERT promoter is highly active in human cancer cells but not in normal differentiated human cells (*6*). MAP kinase signaling pathway is important in regulating hTERT expression and maintenance of tumor cell survival and growth. This pathway provides insight into how tumor cells may escape killing by different agents. In addition, as this pathway is involved in tumor cell survival, several strategies to inhibit the survival pathway are being tested for novel cancer therapies. Inhibition of different MAP kinase signaling, singly, or in particular combination may result in the inhibition of specific event(s) including transcriptional regulation of particular genes such as *hTERT*, which would lead to the inhibition of cancer cell immortalization. As shown in **Fig. 2**, the MAP kinase inhibitor PD98059 inhibits EGF-induced MAP kinase activity and hTERT protein, simultaneously. The inhibition of hTERT protein is apparently at the gene transcription level of hTERT, evidence supported by the findings that PD98059 blocks the hTERT promoter gene transcription activity stimulated by EGF in cultured human breast cancer MCF-7 cells (*see* **Fig. 3**). Although inhibition of the MAP kinase signaling can lead to down-regulation of hTERT expression and telomerase activity, it remains to be determined to what extend telomerase inhibition contributes to cell growth arrest and for how long the telomerase-deficient phenotypes develop while the cell fate may also be regulated by other acute changes because of MAP kinase inhibition.

The inhibition of MAP kinase signaling and *hTERT* gene transcription by PD98059 as shown in **Figs 2** and **3** must intercept the MAP kinase pathway at the site of MEK, interrupting MEK phosphorylation of ERK1/2 kinase, because PD98059 specifically inhibits MEK *(14)*. Thus, many ERK kinase downstream signaling molecules, gene transcription factors, repressors, and modulators are affected by PD98059 inhibition of MEK activation of ERK1/2. Consistently, we have observed that inhibition of MAP kinase signaling by PD98059 in PC-12 cells blocks telomerase down-regulation occurred in association with neural differentiation of the pheochromocytoma cells in the presence of NGF *(7)*. Thus, also as aforementioned, MAP kinase signaling is involved in both up-regulation and down-regulation of *hTERT* gene expression through different downstream elements. We hypothesize that different transcription factors downstream of MAP kinase signaling pathways play different roles in interpreting MAP kinase activation. Transcription factors such as Ets may mediate MAP kinase up-regulation of telomerase by direct binding to Ets consensus site(s) on the *hTERT* gene promoter. Among other over 30 Ets domain transcriptional factors, Elk-1 that is activated by MAP kinase transcriptionally represses c-Myc, suggesting that some Ets family member downstream of MAP kinase activation might be involved in NGF-induced down-regulation of telomerase through repression of Myc expression *(7)*.

Thus, engineering to modify the interface between the *hTERT* gene promoter DNA and its particular key regulator(s) may allow specific control over the *hTERT* gene promoter activity and thereby telomerase activity in cancer. Given that AP-1 binds to *hTERT* gene promoter DNA and inhibits *hTERT* gene transcription *(10)*, the inhibition would then be possibly achieved with small molecules to mimic the action of AP-1 and/or with targeted mutation of *hTERT* gene promoter. In addition, the Ets member ER81 binds *hTERT* gene promoter and induces *hTERT* gene transcription in response to oncogenes that activate MAP kinase signaling *(1)*. Small protein(s) or peptides could be screened to block ER81 stimulation of hTERT transcription. Furthermore, with the recent findings that MAP kinase causes directly phosphorylation at Ser10 of histone 3 leading to *hTERT* gene transcription *(9)*, it is conceivable that targeting the MAP kinase substrate of histone 3 at a particular phosphorylation site on *hTERT* gene promoter may bar MAP kinase execution in histone 3 phosphorylation and *hTERT* gene transcription. In the following sections are detailed the protocols for analyzing the MEK, ERK, JNK, and p38 MAP kinase activities and chemical inhibitors to each MAP kinases pathway. The methods for hTERT promoter reporter assay, and a key downstream target of MAP kinases pathway, are described in detail.

2. Materials
2.1. Cell Culture

1. Stimulant: 50 μg/ml EGF (cat. no. E-9644, Sigma, St. Louis, MO) in phosphate-buffered saline (PBS) containing 0.5 mg/ml bovine serum albumin (BSA) (cat. no. A-9647).
2. Cell medium: Dulbecco's modified Eagle's medium (DMEM) (cat. no. 12430-054, Gibco, Carlsbad, CA) supplemented with 10% fetal calf serum (FCS; cat. no. 12003-500 M, lot no. 1L0403, CSL, Melbourne, Australia). Penicillin/streptomycin (0.5%) (cat. no. 15140-122, Gibco).
3. Sterile Petri dishes (Falcon).

2.2. Protein Extraction

1. Lysis buffer: 20 mM Tris–HCl, pH 7.7, 250 mM NaCl, 2 mM EDTA, 2 mM EGTA, 0.5% NP40, 10% glycerol, 1 mM DTT, 20 mM b-glycerophosphate, 1 mM sodium vanadate, 10 μg/ml leupeptin, 1 mM PMSF, 5 μg/ml aprotinin, 1 μg/ml pepstatin, 1 mM benzamidine, and 100 μM phenanthroline.
2. 2× sodium dodecyl sulfate-polyacrylamide gel electrophoresis (SDS–PAGE) sample buffer: 0.5 M Tris–HCl, 4% SDS, 10% sucrose, 10% 2-mercaptoethanol, and 0.004% bromophenol blue. Make 2 M Tris–HCl (pH 6.8) first, dissolve 242.28 g Tris base in 800 ml H_2O, titrate to pH 6.8 with concentrated HCl, then adjust to 1 l. For 50 ml 2× sample buffer: 12 ml 2 M Tris–HCl (pH 6.8), 2 g SDS, 5 g sucrose, 5 ml 2-mercaptoethanol, and 2 mg bromophenol blue and add H_2O to adjust to 50 ml.

2.3. SDS–PAGE

1. Gel electrophoresis apparatus and power supply.
2. Acrylamide (30%): bisacrylamide (0.8%) solution (cat. no. 161-0158, Bio-Rad, Hercules, CA, USA).
3. Teramethylenediamine (cat. no. 161-0800, Bio-Rad).
4. 10% ammonium persulfate (APS) (cat. no. 161-0700, Bio-Rad).
5. Lower buffer: 1.5 M Tris–HCl, pH 8.8 (separating gel).
6. Upper buffer: 0.5 M Tris–HCl, pH 6.8 (stacking gel).
7. 10% SDS.
8. Running buffer: 25 mM Tris, 192 mM glycine, 0.1% SDS, pH 8.3.

2.4. Western Blotting

1. Immobilon membrane (cat. no. IPVH00010, Millipore, Bedford, MA).
2. 1× TBST: 20 mM Tris–HCl 0.1% Tween 20, 135 mM NaCl, pH 7.6.
3. Make 2 M Tris–HCl (pH 7.6), dissolve 242.28 g Tris base in 800 ml H_2O, titrate to pH 7.6 with concentrated HCl, then adjust to 1 l. For 1× TBST (500 ml), combine 5 ml 2 M Tris–HCl (pH 7.6), 4.383 g NaCl, 0.5 ml Tween 20 into final 494 ml H_2O.
4. 10% fat-free milk powder in 1× TBST.
5. Primary antibody appropriate for MAP kinase signaling.
6. Sigma: general ERK (M5670), ERK1 (M7927), ERK2 (M7556), phos-ERK (M8159), p38 (M0800), phos-p38 (M8177), JNK (J4500), phos-JNK (J4750), MEK (M5795), and phos-MEK (M7683).
7. ERK1 (Sc-93), ERK2 (Sc-154), phos-ERK (Sc-7383), p38 (Sc-535), phos-p38 (Sc-7973), JNK (Sc-474), phos-JNK (Sc-12882), MEK (Sc-219), and phos-MEK (7995) (Santa Cruz, CA, USA).
8. β-tubulin control antibody (cat. no. MAB3408, Chemicon, Temecula, CA).
9. Horseradish peroxidase (HRP)-conjugated secondary antibodies (cat. no. P0161 for anti-mouse and cat. no. P0448 for anti-rabbit, Dako, Carpinteria, CA).
10. Super-Signal (cat. no. 34080, Pierce, Rockford, IL).
11. BioMax autoradiographic film (cat. no. X-OMAT 480 RA, Kodak, Rochester, NY).
12. Measurement of MAP kinase activity: kinase buffer (20 mM HEPES, pH 7.6, 10 mM $MgCl_2$, 10 mM β-glycerophosphate, 0.1 mM Na_3VO_4).

2.5. Inhibitors

See **Table 1** and **Note 1** for inhibitors.

1. MEK inhibitor: PD98059 (2-amino-3-methoxyflavone) (cat. no. 513000, Calbiochem EMD chemicals, Inc. San Diego, USA). Stock solution: 10 mM in dimethylsulfoxide (DMSO) (dissolve 5 mg PD98059 in 1.87 ml DMSO, aliquot into 100 ul/tube, and store at dark and −20°C). PD98059 is a synthetic inhibitor that selectively blocks the activation of MEK1, prevents activation of the MAPKs, ERK1 and ERK2, and inhibits phosphorylation of MAP kinase substrates both in vitro and in vivo *(14,15)*. PD98059 does not inhibit JNK/SAPK and p38 pathways at the concentrations that inhibit ERK activity, specific for ERK pathway *(16,17)*.
2. U0126 is a potent inhibitor for MEK1 and MEK2 kinases. It has significantly higher affinity for all forms of MEK than PD98059, leading to the inhibition of phosphorylation of MEK1/2 and ERK1/2. It also decreases c-Jun expression; c-Fos and c-Jun formed a component of the transcription factor AP-1 *(18–20)*.
3. JNK inhibitor: SP600125 represses c-Jun activation, DNA-binding, and phorbol myristate acetate (PMA)-inducible 92-kDa type IV collagenase expression *(21,22)*.
4. P38 inhibitor: Many of p38 inhibitors have been developed, including SB203580, SB202190, SB242235, SB239063, and SC68376. SB203580 is a pyridinylimidazole

Table 1
MAPK Signal Pathway and Inhibitors

Inhibitors	Compound	Company (Cat.no.)	IC_{50}/Max.	Reference
Inhibitors of MEK1	PD098059* PD184352	Calbiochem (513000) Promega (V1191) Sigma (p-215) BioMol (El-360)	$2\,\mu M/25\,\mu M$	*(14,15)*
Inhibitors of MEK2	U0126* R009- 22110	Calbiochem (662005) Promega (V1221) Sigma (U-120) BioMol (El-282)	$1\,\mu M/5\,\mu M$	*(19,20)*
Inhibitors of ERK1/2	PD098059* E64D Calpeptin	Calbiochem (513000) Promega (V1191) Sigma (p-215) BioMol (El-360)	$2\,\mu M/25\,\mu M$	*(16,17)*
Inhibitors of JNKs	SB600125*	BioMol (El-305)	$0.04\,\mu M/5\,\mu M$	*(22)*
Inhibitors of p38 MAPK	SB203580* SB202190 SB242235	Calbiochem (559389) Promega (V1161) Sigma (S-8307) BioMol (El-286)	$70\,nM/100\,\mu M$	*(24,25)*

Cat. no. only for * compound, more detail will be found at company Web sites.

compound and selective inhibitor of p38 MAP kinase although competitive binding
in the ATP-binding pocket. It has efficacies in several disease models such as
inflammation, septic shock, and injury and also is an inhibitor of caspase 3 activity
and cell death *(23–25)*.

3. Methods

3.1. Cell Culture

Cells are placed in DMEM supplemented with 10% FCS for growth.
Adherent cells are periodically harvested with trypsin–EDTA. In experiments
to test phosphorylation of specific proteins following stimulation and inhibition,
cells were starved in the medium containing 0.1% FCS for 24 h prior to treatments (*see* **Note 2**).

3.2. MAP Kinase Analysis

3.2.1. Biochemical Activity Assay

3.2.1.1. TREATMENTS OF CULTURED CELLS

MAP kinase inhibitors are added to the cells for at least 15 min prior to the addition of the stimulants. The positive control includes NFkB translocation after lipopolysaccharide (LPS). MAP kinase activity can be measured by specific phosphorylation antibodies (*see* **Table 1**) and in vitro protein kinase activity assay using specific substrates.

3.2.1.2. PREPARATION OF WHOLE-CELL LYSATES

1. Cells in the culture plates are placed on ice.
2. Wash three times with ice-cold Dulbecco's PBS.
3. Lyse the cells in ice-cold lysis buffer.
4. After being further solubilized on ice for 30 min, the cellular protein extracts are clarified by centrifugation at $10,000 \times g$ for 10 min at 4°C and frozen at −80°C until use. Total protein concentration in the individual extracts is measured with Bio-Rad protein assay reagents.

3.2.1.3. MEASUREMENT OF MAP KINASE ACTIVITY

1. Equal amounts of cell lysates (10 μg protein) were used to measure MAP kinase activity with rabbit myelin basic protein (MBP) as substrate.
2. Cell lysates were incubated for 20 min at 30°C with 3 μCi [γ-32]-ATP (4500 Ci mmol^{-1}), 50 μM ATP, 10 μg MBP, 10 μg/ml protein kinase inhibitor (PKI), and 50 nM microcystin in 40 μl kinase buffer.
3. After termination of the incubation by addition of 20 μl Laemmli's sample buffer, the phosphorylated proteins were resolved on 15% SDS–PAGE.
4. MBP phosphorylation was visualized by exposure of the dried gel to X-ray film.
5. The activity of MAP kinase was reflected and quantified by MBP phosphorylation, measured with a phosphoimager (Molecular Dynamics Berwyn, Pennsylvania, USA).
6. To verify that MBP phosphorylation thus measured represents MAP kinase activity, we immunoprecipitated MAP kinase in the cell lysates (50 μg protein) with anti-MAP kinase antibody for 1 h at 4°C.
7. Following addition of 20 μl 50% protein A-sepharose 4B suspension for 1 h at 4°C and three washes with ice-cold lysis buffer, the activity of immunoprecipitated MAP kinase was measured as above in 3.2.1.3 step 6 with MBP (0.33 mg/ml) as substrate.
8. Phosphorylated proteins were resolved on 15% SDS–PAGE and transferred onto a nitrocellulose membrane for autoradiography.

9. To control for any variability in protein loading, the membrane was probed with the anti-MAP kinase antibody and MAP kinase protein bands visualized with enhanced chemiluminescence (ECL) detection reagents.

10. After being stripped for 30 min at 50°C in stripping buffer, the membrane was probed once again with an anti-phosphtyrosine antibody to visualize Tyr phosphorylation on MAP kinase.

3.2.2. Western Blotting Analysis

1. Rinse cells with PBS.
2. Lyse in SDS–PAGE sample buffer.
3. Boil 0.1–0.2 mg cell lysates for 15 min, then electrophorese the protein 8–12% SDS–PAGE.
4. Electroblot the separated proteins to Immobilon membrane using the semi-dry transfer method (Bio-Rad).
5. Block the membranes with 10% fat-free milk powder in 1× TBST for 2 h at room temperature.
6. Incubate membranes with appropriate dilutions of specific antibodies in 1× TBST and 10% fat-free milk powder (e.g., anti-ERK or β-tubulin) overnight at 4°C on a rotating wheel.
7. Wash the membranes three times for 10 min with 1× TBST/0.1% Tween 20.
8. Incubate for 1 h at room temperature with appropriate HRP-conjugated secondary antibodies at a 1:1000 dilution in 1× TBST and 10% fat-free milk powder.
9. Wash the membranes another three times for 10 min with 1× TBST/0.1% Tween 20.
10. Use Super-Signal (Pierce) chemiluminescent detection of HRP to visualize the proteins by exposing the membrane to BioMax autoradiographic film. β-tubulin protein expression levels were used to normalize protein loading for each sample (see **Fig. 2**).

3.3. hTERT Gene Promoter Activity Report Assay

The gene reporter assay allows the identification of promoter and enhancers and the study of the correlations between their activities and conformations, by measuring the amount of the reporter protein under the promoter. Gene promoters are upstream of protein coding regions on the genome, regulating expression of the proteins. Reporter gene assay is fast, reliable, cuvet, sensitivity, simple, and high throughput.

3.3.1. Preparation of Cells

Subculture MCF-7 cells for transfection. Seed 0.4×10^5 cells per well in 1 ml DMEM supplemented with 10% FCS into 24-well plates. The cells will be transfected on the next day with 50–70% confluence (see **Note 3**).

To examine the role of MAP kinase signaling in mediating the effect of EGF on *hTERT* gene transcription in the MCF-7 breast cancer cells, we performed *hTERT* gene promoter activity analysis with a *luciferase* gene reporter downstream the *hTERT* gene promoter (*see* **Fig. 1**). If EGF regulates hTERT promoter through MAP kinase signaling, it would be blocked by MAP kinase inhibitors such as PD98059 (*see* **Note 1** and **Fig. 3**). In our protocol, we examined the luciferase report activity under the hTERT 330-bp core promoter. Abolished EGF singling on hTERT 330-bp promoter by mutation of the Ets binding site was used as a negative control. Each assay was performed at least three times in triplicate. Our data and others show that the hTERT 330-bp core promoter that contains one Ets binding site responds to EGF singling. Mutation at –23 bp of the hTERT core promoter or inhibition by MAP kinase inhibitors attenuates the hTERT promoter activity. The Ets transcription factor enhances hTERT expression in immortalized, hTERT-positive cells. Recently, other group reported that *hTERT* gene expression can also be enforced by activating mutations of oncogenes such as HER2/Neu, Ras, and Raf. It is likely that MAP kinase singling through different Ets binding sites located at first exon and intron of hTERT response elements is different from the one we identified within core promoter in different cell types.

3.3.2. Transient Transfection

1. Dilute 1.5–2 µl Fugen transfection reagent for each well in 33 µl DMEM (serum and antibiotic-free medium). Mix well and allow to sit at room temperature for 5 min.
2. With total 1.2 µg DNA (1.0 µg reporter DNA and 0.2 µg β-galactosidase internal control) for each well, combine the DNA with above transfection reagent and DMEM, mix by pipeting up and down 3–5 times, and leave at room temperature for another 15 min to allow formation of DNA–transfection reagent complex (*see* **Note 4**).
3. Add the transfection complex directly into each well, mix well by shaking the cell-culture plate gently, and incubate the cells in a CO_2 incubator for 24 h.
4. After 24 h, replace medium with 500 µl fresh culture medium containing either DMSO (negative control) or 10 µM PD98059. Waiting for 30 min, add EGF to activate MAP kinase pathway, and then analyze luciferase activity in 5–12 h (*see* **Note 5**).

3.3.3. Preparing Cell Lysates and Luciferase Assay

1. Add 4 volumes of water to 1 volume of 5× lysis buffer (cat. no. E3971, Promega, Madison, WI, USA). Equilibrate 1× lysis buffer to room temperature before use (*see* **Note 6**).
2. Carefully remove the growth medium from cells to be assayed, rinse cells with PBS, being careful to not dislodge attached cells, and remove as much of the PBS rinse as possible.

3. Add enough 1× lysis buffer to cover the cells (e.g., 400 μl/6-cm dish, 150 μl/well of a 24-well plate).

4. Rock culture dishes or plates several time to ensure complete coverage of the cells with lysis buffer. Leave these at room temperature for 20 min.

5. Transfer 100 μl cell lysates into 96-well plates, each well contain 100 μl luciferase assay substrate (Luciferase Assay System 10-Pack, cat. no. E1501, Promega). Mix by pipeting two to three times.

6. Place the plate into the luminometer and start reading (*see* **Note 7**).

7. For internal control, β-galactosidase assay is performed by adding 50 μl cell lysates of sample to an equal volume of assay 2× buffer, which contains the substrate *o*-nitrophenyl-beta-D-galactopyranoside (ONPG). Samples are incubated for 30 min, during which time the β-galactosidase hydrolyzes the colorless substrate to *o*-nitrophenyl that is yellow. The reaction is terminated by addition of sodium carbonate, and the absorbance is read at 420 nm with a spectrophotometer (*see* **Note 8**).

8. Analyze the data statistically and graph as shown in **Fig. 2**.

Notes

1. All the inhibitors should be dissolved in sterile DMSO from Sigma. The stock concentrations should be 1000-fold, and the final concentration used to treat cells and solutions should be stored in aliquots at −20°C. Repeated freezing and thawing cycles should be avoided. The inhibitors should be added directly to the medium of cells without further dilution. For control experiments (without inhibitor), the equivalent volume of sterile DMSO should be added to the cells as well. The volume of DMSO should be 0.1% of total volume of medium in which the cells are incubated, as this amount is not toxic. A stock solution of 50 mM PD98059 should be prepared. PD98059 is soluble in aqueous solutions but only up to a concentration of 50 μM. SB203580 and SB202190 stock solution should be prepared at 20 mM.

2. The MAP kinases are very sensitive to any alteration in surrounding conditions, so cell cannot be removed from the incubator or handled in any other way at least 4 h before stimulation.

3. In general, 50–80% confluence will be optimal depending on cell types and transfection reagents, so check the manufacture's manual for special condition.

4. Transfection efficiency varies with cell types and different transfection reagents. β-galactosidase is a commonly used reporter molecule for internal control. The β-Galactosidase Enzyme Assay System with reporter lysis buffer (RLB) (cat. no. E2000) is a convenient method for assaying β-galactosidase activity in lysates prepared from cells transfected with β-galactosidase reporter vectors such as Promega's pSV-β-galactosidase. Green fluorescence protein (GFP) expression vector is another way to monitor the transfection efficiency.

5. Treatment usually occurs after 24 h of transfection. It may need to carry out time course and dose curves for MAP kinase inhibitors, since the duration of

inhibitors varies depending on the reagents used and each target protein and signaling pathway.

6. Promega has three lysis buffers that can be used to prepare cell lysates containing luciferase. Lysis efficiency is dependent on the cell type and needs to be determined for those cells that are resistant to passive lysis. Passive lysis buffer (PLB) will passively lyse cells without the requirement of a freeze-thaw cycle. RLB is mild lysis agent and requires a single freeze-thaw cycle to achieve complete cell lysis. Luciferase Cell Culture Lysis Reagent (CCLR) provides efficient lysis within a few minutes.

7. The plates should be read immediately; the substrate product has a half-life of approximately 10 min.

8. For applications involving coexpression of firefly luciferase with a second reporter gene, we recommend preparing cell lysates with RLB or PLB. CCLR will not yield optimal results when assaying for β-galactosidase or Renilla luciferase coreporter activities.

Acknowledgments

This work was supported by grants from the Australia Research Council and National Health & Medical Research Council of Australia.

References

1. Goueli, B. S., and Janknecht, R. (2004) Upregulation of the catalytic telomerase subunit by the transcription factor ER81 and oncogenic HER2/Neu, Ras, or Raf. *Mol Cell Biol* **24**, 25–35.

2. Maida, Y., Kyo, S., Kanaya, T., Wang, Z., Yatabe, N., Tanaka, M., Nakamura, M., Ohmichi, M., Gotoh, N., Murakami, S., and Inoue, M. (2002) Direct activation of telomerase by EGF through Ets-mediated transactivation of TERT via MAP kinase signaling pathway. *Oncogene* **21**, 4071–9.

3. Bayne, S., and Liu, J. P. (2005) Hormones and growth factors regulate telomerase activity in ageing and cancer. *Mol Cell Endocrinol* **240**, 11–22.

4. Li, H., Xu, D., Toh, B. H., and Liu, J. P. (2006) TGF-â and cancer: is Smad3 a repressor of hTERT gene? *Cell Res* **16**, 169–73.

5. Rajagopalan, H., Bardelli, A., Lengauer, C., Kinzler, K. W., Vogelstein, B., and Velculescu, V. E. (2002) Tumorigenesis: RAF/RAS oncogenes and mismatch-repair status. *Nature* **418**, 934.

6. Liu, J. P. (1999) Studies of the molecular mechanisms in the regulation of telomerase activity. *FASEB J* **13**, 2091–104.

7. Li, H., Pinto, A. R., Duan, W., Li, J., Toh, B. H., and Liu, J. P. (2005) Telomerase down-regulation does not mediate PC12 pheochromocytoma cell differentiation induced by NGF, but requires MAP kinase signalling. *J Neurochem* **95**, 891–901.

8. Alfonso-De Matte, M. Y., Yang, H., Evans, M. S., Cheng, J. Q., and Kruk, P. A. (2002) Telomerase is regulated by c-Jun NH2-terminal kinase in ovarian surface epithelial cells. *Cancer Res* **62**, 4575–8.

9. Ge, Z., Liu, C., Bjorkholm, M., Gruber, A., and Xu, D. (2006) Mitogen-activated protein kinase cascade-mediated histone h3 phosphorylation is critical for telomerase reverse transcriptase expression/telomerase activation induced by proliferation. *Mol Cell Biol* **26**, 230–7.

10. Takakura, M., Kyo, S., Inoue, M., Wright, W. E., and Shay, J. W. (2005) Function of AP-1 in transcription of the telomerase reverse transcriptase gene (TERT) in human and mouse cells. *Mol Cell Biol* **25**, 8037–43.

11. Fu, W., Lu, C., and Mattson, M. P. (2002) Telomerase mediates the cell survival-promoting actions of brain-derived neurotrophic factor and secreted amyloid precursor protein in developing hippocampal neurons. *J Neurosci* **22**, 10710–9.

12. Woo, C. W., Lucarelli, E., and Thiele, C. J. (2004) NGF activation of TrkA decreases N-myc expression via MAPK path leading to a decrease in neuroblastoma cell number. *Oncogene* **23**, 1522–30.

13. Wang, J., Feng, H., Huang, X. Q., Xiang, H., Mao, Y. W., Liu, J. P., Yan, Q., Liu, W. B., Liu, Y., Deng, M., Gong, L., Sun, S., Luo, C., Liu, S. J., Zhang, X. J., and Li, D. W. (2005) Human telomerase reverse transcriptase immortalizes bovine lens epithelial cells and suppresses differentiation through regulation of the ERK signaling pathway. *J Biol Chem* **280**, 22776–87.

14. Alessi, D. R., Cuenda, A., Cohen, P., Dudley, D. T., and Saltiel, A. R. (1995) PD 098059 is a specific inhibitor of the activation of mitogen-activated protein kinase kinase in vitro and in vivo. *J Biol Chem* **270**, 27489–94.

15. Dudley, D. T., Pang, L., Decker, S. J., Bridges, A. J., and Saltiel, A. R. (1995) A synthetic inhibitor of the mitogen-activated protein kinase cascade. *Proc Natl Acad Sci USA* **92**, 7686–9.

16. Johnson, G. L., and Lapadat, R. (2002) Mitogen-activated protein kinase pathways mediated by ERK, JNK, and p38 protein kinases. *Science* **298**, 1911–2.

17. Torres, C., Li, M., Walter, R., and Sierra, F. (2000) Modulation of the ERK pathway of signal transduction by cysteine proteinase inhibitors. *J Cell Biochem* **80**, 11–23.

18. Wityak, J., Hobbs, F. W., Gardner, D. S., Santella, J. B., 3rd, Petraitis, J. J., Sun, J. H., Favata, M. F., Daulerio, A. J., Horiuchi, K. Y., Copeland, R. A., Scherle, P. A., Jaffe, B. D., Trzaskos, J. M., Magolda, R. L., Trainor, G. L., and Duncia, J. V. (2004) Beyond U0126. Dianion chemistry leading to the rapid synthesis of a series of potent MEK inhibitors. *Bioorg Med Chem Lett* **14**, 1483–6.

19. Duncia, J. V., Santella, J. B., 3rd, Higley, C. A., Pitts, W. J., Wityak, J., Frietze, W. E., Rankin, F. W., Sun, J. H., Earl, R. A., Tabaka, A. C., Teleha, C. A., Blom, K. F., Favata, M. F., Manos, E. J., Daulerio, A. J., Stradley, D. A., Horiuchi, K., Copeland, R. A., Scherle, P. A., Trzaskos, J. M., Magolda, R. L., Trainor, G. L., Wexler, R. R., Hobbs, F. W., and Olson, R. E. (1998) MEK

inhibitors: the chemistry and biological activity of U0126, its analogs, and cyclization products. *Bioorg Med Chem Lett* **8**, 2839–44.

20. Ge, X., Fu, Y. M., and Meadows, G. G. (2002) U0126, a mitogen-activated protein kinase kinase inhibitor, inhibits the invasion of human A375 melanoma cells. *Cancer Lett* **179**, 133–40.

21. Shin, M., Yan, C., and Boyd, D. (2002) An inhibitor of c-jun aminoterminal kinase (SP600125) represses c-Jun activation, DNA-binding and PMA-inducible 92-kDa type IV collagenase expression. *Biochim Biophys Acta* **1589**, 311–6.

22. Vincenti, M. P., and Brinckerhoff, C. E. (2001) The potential of signal transduction inhibitors for the treatment of arthritis: is it all just JNK? *J Clin Invest* **108**, 181–3.

23. Cirillo, P. F., Pargellis, C., and Regan, J. (2002) The non-diaryl heterocycle classes of p38 MAP kinase inhibitors. *Curr Top Med Chem* **2**, 1021–35.

24. Ward, K. W., Proksch, J. W., Azzarano, L. M., Salyers, K. L., McSurdy-Freed, J. E., Molnar, T. M., Levy, M. A., and Smith, B. R. (2001) SB-239063, a potent and selective inhibitor of p38 map kinase: preclinical pharmacokinetics and species-specific reversible isomerization. *Pharm Res* **18**, 1336–44.

25. Adams, J. L., Badger, A. M., Kumar, S., and Lee, J. C. (2001) p38 MAP kinase: molecular target for the inhibition of pro-inflammatory cytokines. *Prog Med Chem* **38**, 1–60.

13

Screening of Telomerase Inhibitors

Elke Kleideiter, Kamilla Piotrowska, and Ulrich Klotz

Summary

Shortening of telomeres prevents cells from uncontrolled proliferation. Progressive telomere shortening occurs at each cell division until a critical telomeric length is reached. Telomerase expression is switched off after embryonic differentiation in most normal cells, but it is expressed in a very high percentage of tumors of different origin. Thus, telomerase is regarded as the best tumor marker and a promising novel molecular target for cancer treatment. Therefore, different strategies to inhibit telomerase have been developed. However, systematic screening of telomerase inhibitors has not been performed to compare their therapeutic potential. We propose a suitable strategy for estimation of the therapeutic potential of telomerase inhibitors, which is based on a systematic screening of different inhibitors in the same cell system. From the long list of compounds discussed in the literature, we have selected four telomerase inhibitors of different structure and mode of action: BRACO19 (G-quadruplex-interactive compound), BIBR1532 (non-nucleosidic reverse transcriptase inhibitor), $2'$-O-methyl RNA, and peptide nucleic acids (PNAs; hTR antisense oligonucleotides). To determine minimal effective concentrations for telomerase inhibition, telomerase activity was measured using the cell-free telomerase repeat amplification protocol (TRAP) assay. We also tested inhibitors in long-term cell-culture experiments by exposing A-549 cells to non-cytotoxic concentrations of inhibitors for a period of 99 days. Subsequently, telomerase activity of A-549 cells was investigated using the TRAP assay, and telomere length of samples was assessed by telomere restriction fragment (TRF) Southern blot analysis.

Key Words: Telomerase; telomerase inhibitors; TRAP assay.

From: *Methods in Molecular Biology, vol. 405: Telomerase Inhibition*
Edited by: L. G. Andrews and T. O. Tollefsbol © Humana Press Inc., Totowa, NJ

1. Introduction

Uncontrolled proliferation of tumor cells is probably because of the activity of telomerase present in about 90% of all human cancers *(1)*. As inhibition of telomerase represents a promising approach to anticancer treatment, a number of different experimental and therapeutic strategies have been developed and some of them have been outlined in **Fig. 1**. Proof of principle that inhibition of telomerase can lead to selective anticancer effects is supported by evidence from experiments where inhibition of telomerase results in cell death *(2–4)*. The major components of telomerase, human telomerase reverse transcriptase (hTERT) and human telomerase RNA (hTR), represent possible targets for telomerase inhibitors. hTR antisense oligonucleotides such as peptide nucleic acids (PNAs) and RNA oligomers with substituted methyl-substituted ribose sugar rings (2′-*O*-methyl RNA) prevent telomere elongation by hybridizing to hTR *(5,6)*. Telomerase activity can also be inhibited by direct binding of non-nucleosidic compounds such as BIBR1532 to the catalytic component hTERT *(7)* or by stabilizing the so-called G-quadruplex structure of the single-stranded telomeric end that is accomplished by G-quadruplex-interactive compounds such as BRACO19 *(8)*. As experiments with different telomerase inhibitors have been carried out in different cell models, an exact evaluation of the thera-

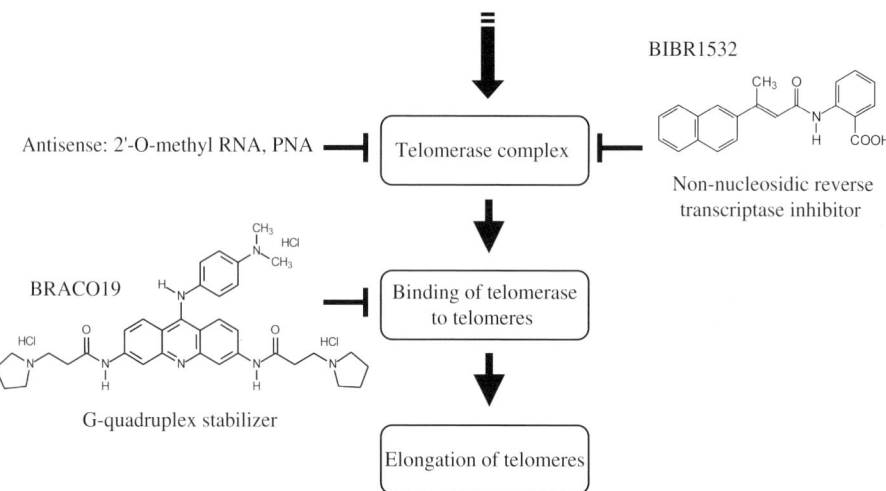

Fig. 1. Selected sites for a pharmacological targeting of telomerase and telomeres (more detailed information on the strategies to inhibit telomerase can be found in other chapters of this book).

peutical potential is difficult to perform. Therefore, we followed an alternative approach to estimate the therapeutic potential of telomerase inhibitors. Our strategy is based on a systematic comparison of the inhibitory action of different compounds on telomerase in a single-cell model.

The majority of cancer-related death is caused by lung cancer *(9)*. Whereas therapeutic management by thoracic surgery can be curative, present options of drug therapy are characterized by low-response rates especially in non-small cell lung cancer (NSCLC). Like for other malignant tumors in different samples of lung cancers, moderate to high telomerase activity could be measured in 60–85% of cases *(10–12)*. These data are in concordance with our preliminary data showing presence of telomerase activity in about 80% (20/25) of tumor samples collected from patients with NSCLC. Thus, agents inhibiting telomerase could offer a new and relative specific therapeutic option for lung cancer treatment. We screened telomerase inhibitors in vitro to evaluate in a comparative manner their therapeutic potential in A-549 cells, an established NSCLC cell line. Screening of telomerase inhibitors starts with the modified cell-free telomerase repeat amplification protocol (TRAP) assay in which we test the inhibitory effect of the substances on telomerase in the cell lysate. Subsequently, to investigate the effect on telomerase activity and telomere length in cell culture, long-term exposure of cells to non-cytotoxic concentrations of inhibitors is performed.

Regarding the first step of our screening, that is, the cell-free TRAP assay, different concentrations of inhibitory compounds were tested and the optimal concentration needed for the inhibition of telomerase was determined. As an example, data for 2′-*O*-methyl RNA are shown in **Fig. 2** The cell-free TRAP assay enables both the determination of the effective inhibitory concentrations of the compounds and the direct comparison of different inhibitors in terms of their inhibitory potency. However, caution is recommended when testing telomerase inhibitors in the TRAP assay. We have shown that inhibitors may interact with components of the reaction mix of the TRAP assay, and this effect is not restricted to one group of substances *(13)*. It is therefore mandatory to test the polymerase chain reaction (PCR) amplification of the TSR8 control oligonucleotide provided in the $TRAP_{EZE}$® kit to assess specific inhibition by the screened substances. It is also recommended to perform the experiments under identical conditions.

After testing telomerase inhibitors in cell lysates, long-term exposure to non-cytotoxic concentrations of inhibitors is performed as the second step of the screening. By estimating the IC_{50} value (the concentration of an inhibitor that is required for 50% inhibition of cell growth) in the 5-bromo-2′-deoxyuridine

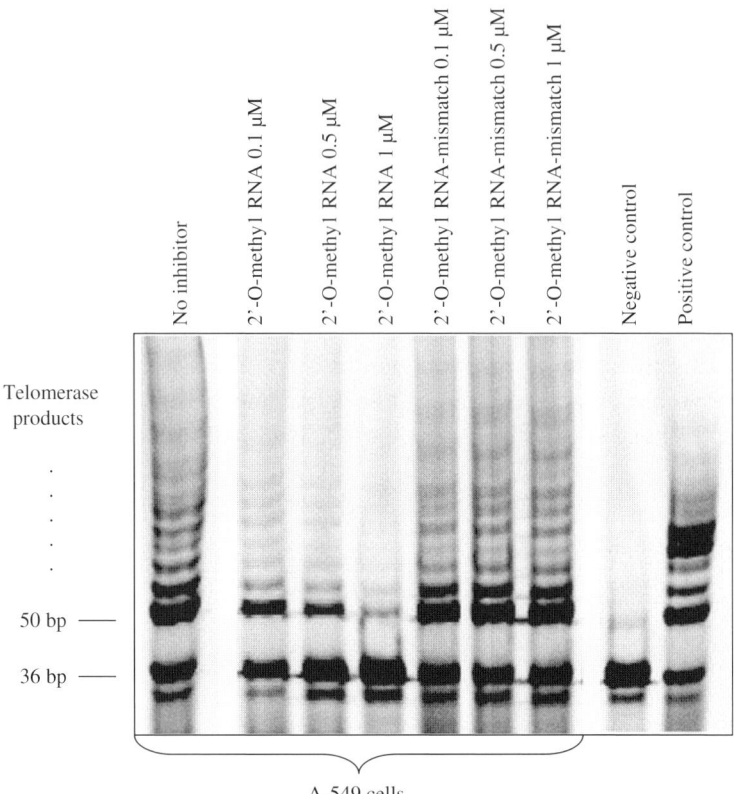

Fig. 2. Telomerase repeat amplification protocol (TRAP) assay showing inhibition of telomerase activity by 2′-O-methyl RNA in A-549 cells. 0.05 μg protein per reaction was used. The lane denoted *No inhibitor* represents control A-549 cells to which no inhibitor was added. Mismatch variant of 2′-O-methyl RNA was applied as a control for sequence specificity of the inhibitor. Negative control: 3-[(3-cholamidopropyl)dimethyl-ammonio]-1-propanesulfonate (CHAPS) lysis buffer. Positive control: TSR8. 36-bp band represents the internal standard.

(BrdU) assay, non-cytotoxic concentrations of the compounds can be selected for their use in cell-culture experiments.

Based on the results of the first step of the screening, we assessed the effect of selected compounds in cell-culture experiments. Telomerase activity and telomere length were determined for A-549 cells treated with inhibitors for 99 days (long-term treatment) using TRAP assay and telomere restriction fragment (TRF) Southern blot analysis. It is necessary to treat cells with the inhibitor for a longer period of time as a low number of cell divisions results only in minor changes in telomere length that cannot be clearly detected by TRF Southern blot analysis.

After completion of testing of telomerase inhibitors in established cell lines, further experiments can be carried out using primary cells obtained from surgically resected tumor material. This would allow for testing the individual susceptibility of the tumor to a particular agent. However, based on our own experience, the number of cell divisions of primary cells is limited; therefore, it will be difficult to perform long-term experiments with those cells. Nevertheless, it appears worthwhile to conduct experiments with short-term incubation, for example, testing synergistic combinations of cytostatics and telomerase inhibitors that could result in a strong enhancement of the anticipated therapeutic effect.

2. Materials

2.1. Cell Culture (A-549 Cells)

1. Roswell Park Memorial Institute (RPMI) 1640 medium (Biochrom, Berlin, Germany) supplemented with 10% fetal calf serum (FCS) (Biochrom), 100 U/ml penicillin, 100 μg/ml streptomycin (Gibco BRL, Karlsruhe, Germany), and 2 mM glutamine (Biochrom).
2. Phosphate-buffered saline (PBS) (Gibco BRL).
3. Trypsin–ethylenediaminetetraacetic acid (EDTA) (Gibco BRL).
4. Trypan blue 0.05% (w/v) (Biochrom).
5. Hemocytometer ("Neubauer") (Roth, Karlsruhe, Germany).

2.2. Determination of Telomerase Activity

2.2.1. Cell Lysis and Estimation of Protein Concentration

1. 3-[(3-cholamidopropyl)dimethyl-ammonio]-1-propanesulfonate (CHAPS) lysis buffer. Store in aliquots at 4 °C.
2. Biciuchoriuic acid (BCA) test (Sigma Aldrich Company, Deisenhofen, Germany). Store at room temperature.
3. Copper (II)-sulfate-pentahydrate 4% (w/v) (Sigma, Deisenhofen, Germany). Store at room temperature.

2.2.2. TRAP

1. TRAP$_{EZE}$® Telomerase Detection Kit (Chemicon International, Hampshire, UK). Store at −20 °C.
2. Oligonucleotide substrate (5'-AATCCGTCGAGCAGAGTT-3') (TS primer, MWG-Biotech AG, Ebersberg, Germany) marked with an infrared dye (Carl Roth

GmbH, Karlsruhe, Germany) to be used instead of the ready-to-use labeled TS primer provided in the TRAP$_{EZE}$® kit to avoid radioactive detection. Store in aliquots in the dark at -20 °C.

3. In some experiments, CX reverse primer (5′-CCCTTACCCTTACCCTTACCCT AA- 3′) has been used instead of the primer mix (provided in the TRAP$_{EZE}$® kit) in the concentration of 30 μM per reaction to assess the inhibitor target specificity (*see* **Note 1**).

4. 10 μl per reaction of solution Q (Qiagen, Hilden, Germany).

2.2.3. Polyacrylamide Gel Electrophoresis

1. 6% Long Ranger Gel (non-denaturating) : Long Ranger Gel solution 50% (Biozym, Hess. Oldendorf, Germany). Store at room temperature. This is a neurotoxin and should be handled carefully, TBE 10× (0.89 M Tris base, 0.89 M boric acid, 10 mM EDTA), distilled water.

2. Ammonium peroxodisulfate solution 10% in distilled water. Store at 4 °C.

3. $N,N,N′,N′$-tetramethyl-ethylenediamine (TEMED, Sigma).

4. Running buffer TBE (0.4×). Store at room temperature.

5. Loading buffer (10×): bromophenol blue 0.41%, xylene cyanol 0.41%, saccharose 67%.

2.3. BrdU Test

1. Cell Proliferation ELISA BrdU (colorimetric) kit (Roche, Mannheim, Germany). Store at 4 °C.

2. ELISA reader "Wallac" (SLT Labinstruments, Crailsheim, Germany).

2.4. Determination of Telomere Length (TRF Analysis)

2.4.1. Genomic DNA Isolation and Digestion of DNA

1. DNA extraction buffer: 10 mM Tris–HCl, 0.1 mM EDTA, 0.5% sodium dodecyl sulfate (SDS), pH 7.5.

2. TE buffer: 10 mM Tris–HCl, 1 mM EDTA, pH 8.0.

3. TBE buffer (10×): 0.89 M Tris base, 0.89 M boric acid, 2 mM EDTA.

4. Loading buffer (10×): 0.41% bromphenol blue, 0.41% xylene cyanol, 67% saccharose.

5. RNase A (Qiagen).

6. Proteinase K (Qiagen).

7. Roti® phenol (Roth).

8. Chloroform (pro analysis) (Merck, Darmstadt, Germany).

9. Agarose ultraPure (Gibco BRL).

10. λ-DNA (Roche).

11. *Hinf*I, NEBuffer 2 (New England Biolabs, Schwalbach, Germany).

2.4.2. Southern Transfer, Hybridization, and Detection

1. Denaturation solution: 1.5 mM NaCl, 0.5 M NaOH.
2. Neutralization solution: 1.0 M Tris base, 1.5 M NaCl.
3. Transfer buffer (20× SSC): 3.0 M NaCl, 0.3 M Tri-Na-citrate, pH 7.0.
4. (Pre)Hybridization mix: 5× SSC, 5% milk powder, 50% formamide, 1% *N*-lauroylsarcosine, 0.02% SDS (telomeric probe 1.25 pmol/ml).
5. Maleic acid buffer: 100 mM maleic acid, 150 mM NaCl, pH 7.5.
6. 10% SDS: SDS, water.
7. 10% blocking buffer: milk powder (Lasana bio skim milk powder, Humana Milchunion eG, Herford, Germany), maleic acid buffer (pH 7.5).
8. Detection buffer: 100 mM Tris–HCl, 100 mM NaCl, 50 mM $MgCl_2$, pH 9.5.
9. Anti-digoxigenin-alkaline phosphatase (AP) (Roche).
10. Disodium3-(4-methoxyspiro{1,2-dioxetane-3,2′-(5′-chloro)tricyclo[3.3.1.1^{3,7}] decan}-4-yl)phenyl phosphate (CSPD®), ready-to-use (Roche).
11. Telomeric probe: (TTAGGG)$_7$, 5′-digoxigenin-label (Roth).
12. 1-kb ladder (DNA-marker) (Gibco BRL; Roche).
13. DIG Oligonucleotide Tailing Kit (Roche).
14. VacuGene XL Vacuum Blotting System (Pharmacia Biotech, Freiburg, Germany).
15. Nylon membranes, positively charged (Roche).
16. The Belly Dancer (Hybridization Water Bath, STOVALL Life Science, Greensboro, NC).
17. Lumi-Film Chemiluminescent Detection Film (Roche).
18. Curix 60 Film Developer (Agfa, Mortsel, Belgium).
19. ONE-DScan® (Version 1994) (Scanalytics, Billerica, Ontario, Canada).
20. Excel® 2002 for Windows (Microsoft Corporation, Munich, Germany).

3. Methods

To determine whether a given compound has a potential to inhibit telomerase, the TRAP assay *(14)* is most often applied. It is a PCR-based method of the detection of telomerase activity in cell lysates. It consists of two steps: (i) addition of a number of telomeric repeats (GGTTAG) by telomerase on the 3′ end of an oligonucleotide substrate (TS) and (ii) amplification of the extended products using the forward TS primer and the reverse RP primer that generates a ladder of products with six base increments detected on a polyacrylamide gel. In case of absence of telomerase or complete telomerase inhibition, no ladder is visible.

In screening potential telomerase inhibitors, it is important to determine the cytotoxic concentration of a given compound. The applied BrdU test is based on the measurement of BrdU incorporation during DNA synthesis in proliferating cells. It enables the determination of the growth-inhibitory effect of different compounds on cell proliferation.

The detection of potential changes in the length of telomeres of treated versus untreated cells is performed using the Southern blot-based telomere length (TRF) analysis.

3.1. Cell Culture

3.1.1. Passaging of A-549 Cells

A-549 cells are passaged twice a week. Every time cells are passaged, number of viable and dead cells are assessed using the vital dye trypan blue and a hemocytometer ("Neubauer"). The reactivity of trypan blue is based on the fact that the chromophore is negatively charged and does not interact with the cell unless the membrane is damaged. Therefore, all cells that exclude the dye are viable.

Dependent on the growth kinetics, a defined number of A-549 cells are seeded at each passaging to ensure that cells can be passaged twice a week.

After seeding, 2 h-incubation is allowed for the cells to adhere and a fresh aliquot of inhibitor solution is added.

3.2. Determination of Telomerase Activity in Cell Lysates

3.2.1. Preparation of Samples

Preparation of the samples is carried out according to the $TRAP_{EZE}$® kit protocol. All steps are carried out on ice.

1. 1×10^6 cells are resuspended in $200\,\mu l$ CHAPS lysis buffer. For a different number of cells, change appropriately the buffer volume (*see* **Note 2**).
2. Samples are incubated on ice for 30 min.
3. Samples are centrifuged at $\simeq 15,300\,g$ for 20 min at 4 °C.
4. Two aliquots are taken from the supernatant: one for the determination of the protein concentration and one for the PCR (*see* **Note 3**).
5. Protein concentrations are measured using the BCA test. Using lysis buffer, bovine serum albumin (BSA) standard dilutions ($0–50\,\mu g$) and a dilution series of the extract in a total volume of $50\,\mu l$ are prepared. $5–10\,\mu l$ extract is a typical range. 1 ml Protein Assay Reagent is added to each standard or sample. Reaction mixtures have to be mixed well and incubated for 30 min at 37 °C. Absorbance is read at OD_{562} within 30–60 min. Determination of extract protein concentration will be assessed from BSA standard plot of OD_{562} versus µg BSA ($0–24\,\mu g/ml$). If the sample volume does not allow to perform the BCA test in a total volume of $50\,\mu l$, the determination can also be performed with smaller sample volumes in microtiter plates.
6. The rest of the samples is frozen in liquid nitrogen and stored at -80 °C.

3.2.2. TRAP Assay

All steps are carried out on ice. It is advisable to use different sets of pipettes for sample and master mix preparation. Moreover, it is recommended to prepare the master mix and samples in separate rooms.

1. After thawing the TRAP$_{EZE}$® kit, the master mix solution is prepared (*see* **Note 4**). 5 pmol TS primer and 10 µl solution Q are used per reaction, and the telomerase inhibitor is added directly to the master mix (the volume of solution Q and the inhibitor is subtracted from the volume of water). 2 µl sample is added to the reaction mixture (*see* **Note 5**).
2. 2 µl TSR8 control template and 2 µl CHAPS lysis buffer are used as a positive and a negative control, respectively (*see* **Note 6**).
3. The samples are placed in a thermocycler and subjected to the following conditions: (i) incubation at 30 °C for 30 min and telomerase inactivation at 94 °C for 2 min, (ii) 35 cycles of denaturation at 94 °C for 2 min and synthesis at 57 °C for 30 s, a single post-synthesis step at 57 °C for 30 s. The end temperature is set at 4 °C.
4. The PCR products are separated vertically on the 6% polyacrylamide gel (0.4-mm thick) using a sequencer. For an approximately 470 cm² gel, the following compounds are mixed: 33.5 ml water, 4.9 ml 50% acrylamide solution, 1.6 ml 10× TBE solution, 327 µl 10% APS, 40 µl TEMED. The gel should be left for 1.5–2 h to polymerize. Running conditions are as follows: 35 mA, 31.5 W, 1500 V at 50 °C (*see* **Note 7**).
5. Telomerase activity is estimated visually by taking the number of bands and their intensity into account. The lower the number of bands on the gel, the stronger the telomerase inhibition.

3.3. BrdU Assay

The assay is performed using the Cell Proliferation ELISA BrdU kit. The Cell Proliferation ELISA BrdU kit is designed to quantitate cell proliferation based on the measurement of BrdU incorporation during DNA synthesis in proliferating cells. Thus, to determine the cytotoxic and non-cytotoxic concentrations of a given compound, cells are incubated with different concentrations of the respective substance. The BrdU assay enables the determination of the growth-inhibitory effect (e.g., *IC*$_{50}$ value) of compounds on cell proliferation. Based on these data, non-cytotoxic concentrations of the compounds can be selected for their further use in cell-culture experiments.

The BrdU assay is performed according to the manufacturer's instruction manual.

A-549 cells are seeded in a flat-bottomed 96-well plate (5000 cells/well). After the cells become adherent, the respective concentrations of telomerase inhibitors are added and cells are cultured in the presence of the test substances in a 96-well microtiter plate at 37 °C for 48 h. Subsequently, BrdU is added to the cells, and the cells are reincubated for 24 h. During this labeling period, the pyrimidine analog BrdU is incorporated in place of thymidine into the DNA of proliferating cells. After removing the culture medium, cells are fixed and the DNA is denatured (denaturation of the DNA is necessary to improve the accessibility of the incorporated BrdU for detection by the antibody). The anti-BrdU-peroxidase (POD) binds to the BrdU incorporated in newly synthesized, cellular DNA. The immune complexes are detected by the subsequent substrate reaction. The reaction product is quantified by measuring the absorbance at the respective wavelength (405 nm) using a scanning multiwell spectrophotometer (ELISA reader). The developed color and thereby the absorbance values directly correlate to the amount of DNA synthesis and to the number of proliferating cells in the respective microcultures. IC_{50} values are determined on the basis of cell-growth profiles.

Limitations to the interpretation of results regarding cytotoxicity are recognized as the BrdU assay can not differentiate between cell senescence or cell death.

3.4. Long-Term Treatment with Inhibitor

Long-term treatment of cells with an inhibitor is carried out for up to 99 days. After each passage, cells are incubated for about 2 h to allow their adherence to the bottom of the flask and subsequently a fresh solution of the inhibitor is directly added to the medium. At regular time intervals, cell pellets are frozen in liquid nitrogen and stored at −80 °C for further analysis (1×10^6 cells for TRAP and 5×10^6 cells for TRF; *see* **Note 8**).

3.5. Determination of Telomere Length (TRF Analysis)

Unless noted all incubation steps were performed at room temperature.

3.5.1. Genomic DNA Isolation and Digestion of DNA

1. Up to 5×10^6 cells are resuspended in extraction buffer (0.5 ml buffer/10^6 cells) and 40 μl RNase A (10 mg/ml) are added to each sample according to the QIAamp® DNA Mini Kit protocol.
2. After 60-min incubation at 37 °C with RNase A, 6 μl Proteinase K (20 mg/ml) is added per 1 ml cell extract and samples are incubated at 50 °C overnight.
3. DNA is isolated from cell pellets using the standard protocol of phenol–chloroform extraction (*see* **Note 9**). Ethanol-precipitated DNA is resuspended in TE buffer and quantified spectrophotometrically. Integrity of DNA is monitored by conventional electrophoresis in 0.8% agarose gels. TRFs are obtained from genomic DNA by

digestion using the restriction enzyme *Hinf*I (add buffer 2 and sterile water to the DNA, mix, and add the enzyme). Incubate the samples overnight at 37 °C. Successful digestion of DNA is monitored by agarose gel electrophoresis (0.8%).

3.5.2. Southern Blotting and Hybridization

1. Up to 15 μg cleaved DNA is separated in 0.8% agarose gels (∼ 2 V/cm for 15–16 h). The 1-kb DNA ladder serves as base pair marker.
2. The gel is stained with ethidium bromide (0.5 μg/ml in 1× TBE), and fragments are detected under UV light (302 nm).
3. Depurinization is performed by UV irradiation (302 nm) for 5 min.
4. DNA is transferred to the nylon membrane using a vacuum blotting system. After 20-min denaturation at 50 mbar, a neutralization step is carried out (20 min, 50 mbar) and DNA transfer is performed with transfer buffer (20× SSC) for 90 min at 50 mbar.
5. Membranes are washed with 2× SSC for 5 min, and DNA is cross-linked by UV irradiation (254 nm, 124 mJ; *see* **Note 10**).
6. Prehybridization is performed for at least 30 min at 42 °C using a hybridization oven. The hybridization step is carried out overnight at 42 °C. In contrast to the prehybridization buffer, the hybridization buffer contains the telomeric probe. The telomeric probe is enzymatically labeled at the 3′ end with digoxigenin using the "DIG Oligonucleotide Tailing Kit."
7. Membranes are washed twice for 15 min with 2× SSC/0.1% SDS at 42 °C. Two washing steps in washing buffer follow (in each case 5 min).

3.5.3. Detection Procedure

Detection of AP is based on the formation of a chemiluminescent signal. CSPD® serves as substrate, which is dephosphorylated by AP. Enzymatic dephosphorylation of CSPD® leads to a metastable phenolate anion that decomposes and emits light at a maximum wavelength of 477 nm. The luminescent light emission is recorded on X-ray films.

1. To block unspecific binding sites, membranes are incubated for at least 1 h in blocking solution (*see* **Note 11**).
2. Antibody (anti-digoxigenin-AP) is diluted 1:5000 in blocking solution, and the membranes are incubated with the antibody solution for at least 45 min.
3. To remove unbound antibody, membranes are washed twice in maleic acid/0.1% Tween® 20 for 15 min.
4. For the development of the chemiluminescent signal, membranes are equilibrated with the detection buffer (pH 9.5) and placed in hybridization bags with 1 ml CSPD®. Membranes are incubated for 5 min at room temperature in the dark.
5. Excess liquid is removed, and membranes are incubated for at least 30 min at 37 °C to enhance the luminescent reaction.

6. After development, the films are scanned and analyzed using One-Dscan 1.0 and Excel 2002 software (troubleshooting, *see* **Note 11**). Mean TRF length is calculated as weighted mean of the optical density (OD) using the formula $L = \Sigma(OD_i \times L_i)/\Sigma OD_i$, with OD_i as the integrated signal in interval i and L_i as the TRF length at the midpoint of interval i. Average telomere length is determined at least in duplicate whenever sufficient DNA is available.

4. Notes

1. It has been shown that telomerase inhibitors may interact with components of the primer mix provided in the TRAP$_{EZE}$® kit *(13)*. It is therefore mandatory to test the PCR amplification of the TSR8 control oligonucleotide provided in the TRAP$_{EZE}$® kit to assess the specificity of the inhibitors. The test of specificity can be addressed by modifying the analytical conditions in such a way that the primer mix provided in the TRAP$_{EZE}$® kit is replaced by a different reverse primer (CX). In contrast to the unknown reverse primer RP from the TRAP$_{EZE}$® kit, sequence and concentration of the CX primer are known. For details, see **ref. *13***.
2. It is possible to use frozen cells. In this case, leave the frozen cell pellet on ice, wait until it thaws, and apply CHAPS lysis buffer as described.
3. The samples should be frozen in liquid nitrogen and stored at −80 °C until further processing.
4. Larger amounts of master mix can be prepared and stored in aliquots for several weeks at −20 °C.
5. For A-549 cells, optimal results are achieved with 0.05 µg protein per reaction.
6. To obtain a clearer ladder pattern on the gel, TSR8 can be diluted 1:2 with water. We have previously shown *(13)* that the positive control TSR8 that is provided in the TRAP kit can be also inhibited by some inhibitors; therefore, we recommend that an additional sample of TSR8 with the inhibitor to be tested should be always included in the TRAP assay.
7. APS and TEMED should be added at the end, as they facilitate the polymerization of the gel. Detection of PCR products using a sequencer allows separation of PCR products on 6% polyacrylamid gels (0.4-mm thick). As the PCR products are detected using the infrared dye-labeled TS primer, staining of products by SYBR® Green/ethidium bromide or radioactive detection is avoided. However, the described method differs from the most common methods, which uses vertical 10–12% polyacrylamid gels and thicker gels. Further details regarding the detection of the PCR products are described in the manufacturer's instruction manual of the TRAP$_{EZE}$® kit.
8. It is recommended to test aliquots of the inhibitor in the cell-free TRAP assay at the beginning, in the middle, and at the end of long-term experiments to prove the biological activity of the compound. Stable storage conditions should be established for each inhibitor tested.

9. The standard phenol–chloroform extraction enables obtaining high amounts of DNA that is necessary for Southern blot analysis.
10. Membranes should be dried before cross-linking.
11. Blocking solution should be prepared always fresh. As the TRF analysis depends on the quality of developed and scanned X-ray films, the films and scans should be of high quality. To ensure that peaks can easily be identified, development of X-ray films should vary regarding exposition time and films should be scanned using a transmitted light scanner.

Acknowledgments

We thank Jessica Lauser for her excellent technical assistance. Tumor material was kindly provided by Dr G. Friedel from the Department of Thorax Surgery, Schillerhöhe Clinical Center, Gerlingen, Germany, and Dr P. Fritz from Robert Bosch Hospital in Stuttgart, Germany. The project is financially supported by German Cancer Aid, Bonn, Germany, and Robert Bosch Foundation, Stuttgart, Germany.

References

1. Janknecht, R. (2004) On the road to immortality: hTERT upregulation in cancer cells, *FEBS Lett.* **564**, 9–13.
2. Nakajima, A., Tauchi, T., Sashida, G., Sumi, M., Abe, K., Yamamoto, K., Ohyashiki, J. H., and Ohyashiki, K. (2003) Telomerase inhibition enhances apoptosis in human acute leukemia cells: possibility of antitelomerase therapy, *Leukemia* **17**, 560–567.
3. Shammas, M. A., Shmookler Reis, R. J., Li, C., Koley, H., Hurley, L. H., Anderson, K. C., and Munshi, N. C. (2004) Telomerase inhibition and cell growth arrest after telomestatin treatment in multiple myeloma, *Clin. Cancer Res.* **10**, 770–776.
4. Shammas, M. A., Koley, H., Batchu, R. B., Bertheau, R. C., Protopopov, A., Munshi, N. C., and Goyal, R. K. (2005) Telomerase inhibition by siRNA causes senescence and apoptosis in Barrett's adenocarcinoma cells: mechanism and therapeutic potential, *Mol. Cancer* **4**, 24.
5. Norton, J. C., Piatyszek, M. A., Wright, W. E., Shay, J. W., and Corey, D. R. (1996) Inhibition of human telomerase activity by peptide nucleic acids, *Nat. Biotechnol.* **14**, 615–619.
6. Pitts, A. E. and Corey, D. R. (1998) Inhibition of human telomerase by 2'-O-methyl-RNA, *Proc. Natl. Acad. Sci. U. S. A.* **95**, 11549–11554.
7. Damm, K., Hemmann, U., Garin-Chesa, P., Hauel, N., Kauffmann, I., Priepke, H., Niestroj, C., Daiber, C., Enenkel, B., Guilliard, B., Lauritsch, I., Muller, E., Pascolo, E., Sauter, G., Pantic, M., Martens, U. M., Wenz, C., Lingner, J., Kraut, N.,

Rettig, W. J., and Schnapp, A. (2001) A highly selective telomerase inhibitor limiting human cancer cell proliferation, *EMBO J.* **20**, 6958–6968.

8. Gowan, S. M., Harrison, J. R., Patterson, L., Valenti, M., Read, M. A., Neidle, S., and Kelland, L. R. (2002) A G-quadruplex-interactive potent small-molecule inhibitor of telomerase exhibiting in vitro and in vivo antitumor activity, *Mol. Pharmacol.* **61**, 1154–1162.

9. Landis, S. H., Murray, T., Bolden, S., and Wingo, P. A. (1999) Cancer statistics, 1999, *CA Cancer J. Clin.* **49**, 8–31.

10. Hsu, C. P., Miaw, J., Hsia, J. Y., Shai, S. E., and Chen, C. Y. (2003) Concordant expression of the telomerase-associated genes in non-small cell lung cancer, *Eur. J. Surg. Oncol.* **29**, 594–599.

11. Dikmen, E., Kara, M., Dikmen, G., Cakmak, H., and Dogan, P. (2003) Detection of telomerase activity in bronchial lavage as an adjunct to cytological diagnosis in lung cancer, *Eur. J. Cardiothorac. Surg.* **23**, 194–199.

12. Albanell, J., Lonardo, F., Rusch, V., Engelhardt, M., Langenfeld, J., Han, W., Klimstra, D., Venkatraman, E., Moore, M. A., and Dmitrovsky, E. (1997) High telomerase activity in primary lung cancers: association with increased cell proliferation rates and advanced pathologic stage, *J. Natl. Cancer Inst.* **89**, 1609–1615.

13. Piotrowska, K., Kleideiter, E., Mürdter, T. E., Tätz, S., Baldes, C., Schäfer, U., Lehr, C.-M., and Klotz, U. (2005) Optimization of the TRAP assay to evaluate specificity of telomerase inhibitors. *Lab. Invest.* **85**, 1565–1569.

14. Kim, N. W., Piatyszek, M. A., Prowse, K. R., Harley, C. B., West, M. D., Ho, P. L., Coviello, G. M., Wright, W. E., Weinrich, S. L., and Shay, J. W. (1994) Specific association of human telomerase activity with immortal cells and cancer, *Science* **266**, 2011–2015.

14

Telomerase Inhibition Combined with Other Chemotherapeutic Reagents to Enhance Anti-Cancer Effect

Tetsuzo Tauchi, Junko H. Ohyashiki, and Kazuma Ohyashiki

Summary

Genetic experiments using a dominant-negative form of human telomerase (DN-hTERT) demonstrated that telomerase inhibition can result in telomeric shortening followed by proliferation arrest and cell death by apoptosis. Neoplastic cells from telomerase RNA null (mTERC$^{-/-}$) mice showed enhanced chemosensitivity to doxorubicin or related double-strand DNA break (DSB)-inducing agents. Telomerase dysfunction, rather than telomerase inhibition, is proposed to be the principal determinant governing chemosensitivity specifically to DSB-inducing agents. We observed that imatinib and vincristine (VCR), in addition to DSB-inducing agents, also enhanced chemosensitivity in telomestatin-treated K562 cells. This observation suggests that combined use of telomerase inhibitors and imatinib or other chemotherapeutic agents may be a very useful approach to treatment of BCR-ABL-positive leukemia.

Key Words: Telomerase inhibition; G-quadruplex; apoptosis; hTERT.

1. Introduction

There has been an increase in telomere–telomerase research over the past ten years, with many different pre-clinical approaches being tested for inhibiting the activity of telomerase as a novel therapeutic modality for the treatment of cancers. In this review, we will provide some basic background information about telomeres and telomerase and then discuss the challenges of approaches that are currently under investigation.

From: *Methods in Molecular Biology, vol. 405: Telomerase Inhibition*
Edited by: L. G. Andrews and T. O. Tollefsbol © Humana Press Inc., Totowa, NJ

2. Some Concerns About Anti-Telomerase Therapy for Malignancies

Telomerase is a cellular RNA-dependent DNA polymerase that serves to maintain the tandem arrays of telomeric TTAGGG repeats at eukaryotic chromosome ends (1). Telomeres are highly conserved in organisms ranging from unicellular eukaryocytes to mammals, indicating a strong role of their protective mechanisms in preventing chromosomal ends from undergoing degradation and ligation with other chromosomes (1). Without telomeric caps, human chromosomes would undergo end-to-end fusions, with formation of dicentric and multicentric chromosomes (2). These abnormal chromosomes would break during mitosis, resulting in severe damage to the genome and activation of DNA damage checkpoints, leading to cell senescence or initiation of the apoptosis cell death pathway (3). Indeed, it has been proposed that telomere length specifies the number of cell divisions a cell can undergo before senescence (4). Telomerase activity is upregulated in the vast majority of human tumors as compared with normal somatic tissues. Expression of the catalytic subunit of telomerase, hTERT, in cultured human primary cells reconstitutes telomerase activity and allows immortal growth (5–8). TERT-mediated telomerase activation is able to cooperate with oncogenes in transforming cultured primary cells into neoplastic cells (9). In addition, it has been shown that TERT-derived cell proliferation results in activation of the c-myc oncogene (10). These findings in cultured cells have opened up the possibility that telomerase upregulation may contribute actively to tumor growth (11). As a corollary to this hypothesis, the inhibition of telomerase in tumor cells should disrupt telomere maintenance and turn malignant cells toward proliferative crisis, followed by senescence or cell death. Genetic experiments using a dominant-negative form of human telomerase (DN-hTERT) have demonstrated that telomerase inhibition can result in telomere shortening followed by proliferation arrest and cell death by apoptosis (12,13). This makes telomerase a target not only for cancer diagnosis but also for the development of novel therapeutic agents.

There are several considerations about telomerase as an anticancer target that need to be addressed. First, there will be an expected lag phase between the time telomerase is inhibited and the time when the telomeres of cancer cells shorten sufficiently to produce detrimental effects on cellular proliferation. This lag phase will vary depending on the initial telomere length. There is, at least in theory, the possibility that cancer cells might become resistant to telomerase inhibitors or develop alternative mechanisms of telomere maintenance independent of telomerase, which has been seen in experimentally immortalized human cell lines. Finally, inhibitors of telomerase would potentially have effects on other human somatic cells that express telomerase, such as hematopoietic stem cells, germline cells, and cells of the basal layer of the epidermis and intestinal crypts. We believe that these effects might be minor because stem cells of renewal tissues typically have much longer telomeres than cancer cells have and the deepest stem cells only proliferate intermittently. During the time that these cells are quiescent, telomere shortening does not occur and telomerase activity is negligible.

There have also been concerns that inhibiting telomerase might lead to an increase in malignancy by enhancing the genomic instability of cells *(14)*. These concerns have arisen from observations of $mTR^{-/-}$ knockout mice, which have an increased incidence of malignancies in both early and late generations, particularly in the setting of p53 mutant tumors *(14)*. There is no evidence to suggest that this would be the case in humans *(15)*. Mouse telomeres are much longer than even the longest human telomeres and do not appear to have a role in signaling senescence.

3. Telomerase Inhibition with Other Chemotherapeutic Reagents to Enhance Anti-Cancer Effect

Would inhibition of the enzyme in the classical sense, which would require a lag phase before any detrimental effects on the cells, be a reasonable strategy in patients with a large tumor burden? The use of telomerase inhibitors in this situation would be an adjuvant treatment in combination with chemotherapeutic reagents as well as new reagents such as tyrosine kinase inhibitors.

We have previously demonstrated the enhancement of sensitivity to imatinib in DN-hTERT-expressing K562 cells. To assess the effects of telomerase inhibition in modulating responses to imatinib, which is a selective inhibitor of BCR-ABL tyrosine kinase, experiments have focused on early-passaged K562 cells (PD10) expressing DN-hTERT and wild-type hTERT (WT-hTERT) (*see* **Fig. 1A**). In a series of experiments, pools of K562 cells that expressed DN-hTERT or WT-hTERT were cultured with imatinib for 48 h (*see* **Fig. 1A**). The incidence of apoptosis was determined by flow-cytometric analysis with APO2.7 monoclonal antibody (mAb) (*see* **Fig. 1A**). DN-hTERT-expressing K562 cells showed enhanced induction of apoptosis compared with WT-hTERT-expressing cells on exposure to 0.5 or 1 µM imatinib (*see* **Fig. 1A**). Although K562 cells expressing DN-hTERT at PD20 showed features of apoptosis spontaneously, DN-hTERT-expressing K562 cells at PD10 showed no difference in the features of apoptosis without imatinib. These results suggest that there is a cytotoxic synergy between telomerase inhibition and imatinib. To elucidate the mechanism of a cytotoxic synergy between telomerase inhibition and imatinib, we examined the effect

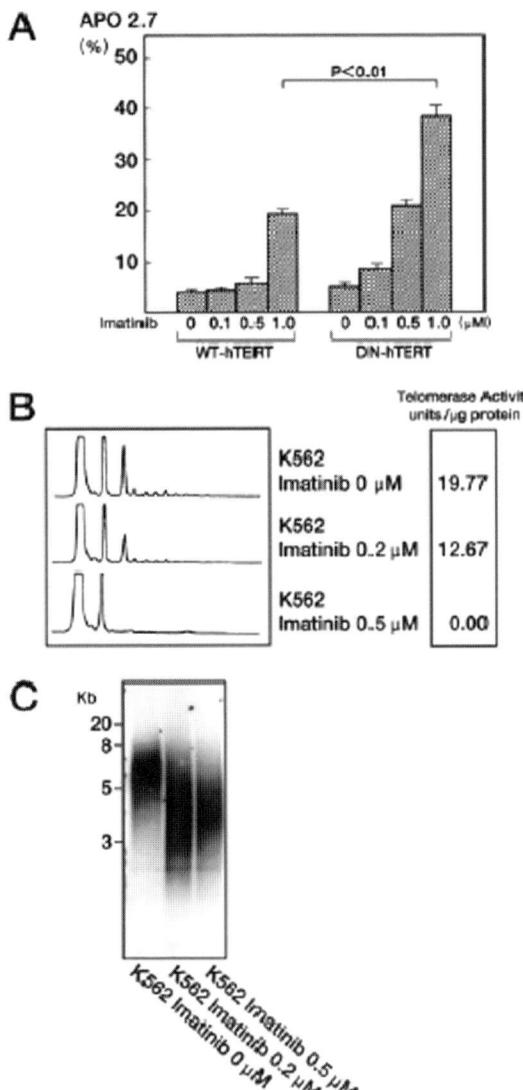

Fig. 1. Enhancement of sensitivity to imatinib in DN-hTERT-expressing K562 cells. (**A**) Either DN-hTERT-expressing (C3, C6, and C7) or WT-hTERT-expressing (C1–C3) K562 cells at PD10 were cultured with the indicated concentration of imatinib for 48 h. The incidence of apoptosis was determined by flow-cytometric analysis with APO2.7 monoclonal antibody (mAb). Values are shown as mean ± SD of triplicates. Similar results were obtained in two independent experiments. (**B**) K562 cells were incubated with the indicated concentrations of imatinib for 48 h, then telomerase activity

of imatinib on telomerase activity and telomere length in K562 cells (*see* **Fig. 1B** and **C**). K562 cells were incubated with imatinib for 48 h, then we investigated the telomerase activity in these cells by telomere repeat amplification protocol (TRAP) assay (*see* **Fig. 1B**). 0.2 or 0.5 µM of imatinib had no effect on short-term cell viability. Imatinib dramatically reduced the telomerase activity in K562 cells (*see* **Fig. 1B**). To examine the effect of imatinib on telomere length in K562 cells, K562 cells were incubated with the indicated concentration of imatinib for 10 days, then telomere length were examined by Southern blotting (*see* **Fig. 1C**). The average telomere restriction fragment size from K562 cells was shortened progressively (*see* **Fig. 1C**). These results suggest that imatinib regulates the telomerase activity and telomere length in BCR-ABL-transformed cells.

The enhanced sensitivity to imatinib may imply that there is cytotoxic synergy between telomerase inhibition and imatinib. Recently, Maida et al. *(16)* reported that epidermal growth factor activates telomerase through Ras/MEK/ERK pathways. We observed that imatinib suppressed telomerase activity and shortened terminal restruction fragments (TRF) in K562 cells (*see* **Fig. 1B** and **C**). In this regard, imatinib may suppress telomerase activity through the inhibition of Ras pathways. It has been shown that activation of the nuclear c-Abl protein can contribute to the induction of apoptosis *(17)*. Actually, Kharbana et al.*(18)* have reported that c-Abl protein is directly associated with hTERT and inhibits telomerase activity. As imatinib stimulates nuclear import of BCR-ABL by combining with leptomycin B *(19)*, imatinib may affect subcellular localization of BCR-ABL in the situation of telomere dysfunction. Although the exact mechanism of this apoptotic effect in DN-hTERT-expressing cells requires further elucidation, our observation suggests that certain anti-neoplastic agents may enhance the telomere position effect in neoplastic cells *(20)*. We observed only the enhanced induction of apoptosis by imatinib in K562 cells expressing DN-hTERT (*see* **Fig. 1**), but inhibitors that are combined with the use of imatinib seem to be very useful for BCR-ABL-positive leukemia.

We also examined the impact of telomerase inhibition by telomestatin on chemotherapeutic responses (*see* **Fig. 2A–D**). Telomestatin is a natural product

---

Fig. 1. was examined by a telomere repeat amplification protocol (TRAP) assay. (C) K562 cells were cultured with the indicated concentrations of imatinib for 10 days, then total genomic DNA from the indicated cells was assessed for telomere restriction fragment size by Southern blot analysis with a telomeric probe.

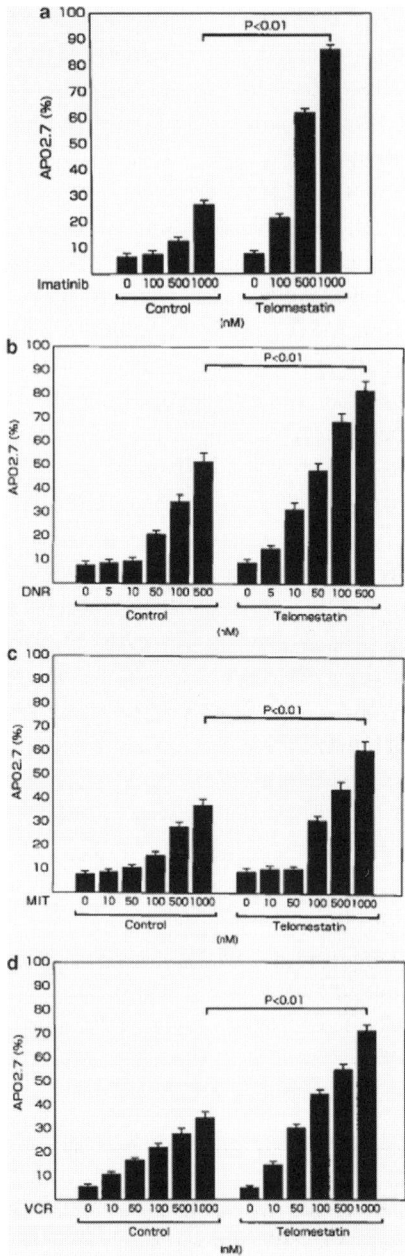

Fig. 2. Enhancement of apoptosis in telomestatin-treated K562 cells by chemotherapeutic agents. K562 cells were cultured with $2 \mu M$ telomestatin for 10 days, and

isolated from *Streptomyces anulatus* 3533-SV4 and has been shown to be a very potent telomerase inhibitor *(21)*. The structural similarity between telomestatin and a G-quadruplex suggested that the telomerase inhibition may be because of its ability either to facilitate the formation of or to trap out preformed G-quadruplex structures and thereby sequester single-stranded d[T2AG3]n primer molecules required for telomerase activity *(21)*. In fact, telomestatin selectively facilitates the formation of, or stabilizes, intramolecular G-quadruplexes, including that produced from the human telomeric sequence d[T2AG3]4 *(21)*. Mechanistically distinct classes of reagents were selected for analysis, including imatinib, daunorubicin (DNR), mitoxantrone (MIT), and vincristine (VCR). To assess the effects of telomerase inhibition in modulating responses to these reagents, experiments focused on early passaged telomestatin-treated K562 cells (PD10). In this series of experiments, K562 cells were cultured with telomestatin for 10 days, subsequently the telomestatin-treated K562 cells were incubated with the agents for 72 h, and the incidence of apoptosis was determined by flow-cytometric analysis with APO2.7 mAb (*see* **Fig. 2A–D**). The telomestatin-treated K562 cells showed enhanced induction of apoptosis compared with control cells after exposure to imatinib, DNR, MIT, and VCR (*see* **Fig. 2A–D**), whereas significant chemosensitivity was not observed in cells exposed to VP-16, 6-MP, and methatrexate (MTX). These results, demonstrating enhanced sensitivity to some classes of chemotherapeutic agents, imply cytotoxic synergy between telomere dysfunction and these agents. We conclude that telomerase inhibitors combined use of imatinib and other chemotherapeutic agents may be very useful for the treatment of human leukemia.

References

1. Blackburn, E., and Greider, C. (eds) *Telomeres*. New York: Cold Spring Harbor Laboratory Press, 1995.
2. van Steensel, B., Smogorzewska, A., and de Lange, T. TRF protects human telomeres from end-to-end fusions. *Cell*, *92*: 401–413, 1998.
3. de Lange, T., and Jacks, T. For better or worse? Telomerase inhibition and cancer. *Cell*, *98*: 273–275, 1999.

Fig. 2. subsequently these cells were incubated with the indicated concentrations of imatinib (**A**), daunorubicin (DNR) (**B**), mitoxantrone (MIT) (**C**), and vincristine (VCR) (**D**) for 72 h. The incidence of apoptosis was determined by flow-cytometric analysis with APO2.7 monoclonal antibody (mAb).

4. Harley, C.B. Telomere loss: mitotic clock or genetic time bomb? *Mutat. Res.*, *256*: 271–282, 1991.

5. Bodnar, A.G., Ouellette, M., Frolkis, M., Holt, S.E., Chiu, C.P., Morin, G.B., Harley, C.B., Shay, J.W., Lichtsteiner, S., and Wright, W.E. Extension of life-span by introduction of telomerase into normal human cells. *Science*, *279*: 349–352, 1998.

6. Kiyono, T., Foster, S.A., Koop, J.I., McDougall, J.K., Galloway, D.A., and Klingelhutz, A.J. Both Rb/p16INK4a inactivation and telomerase activity are required to immortalize human epithelial cells. *Nature*, *396*: 84–88, 1998.

7. Jiang, X.R., Jimenetz, G., Chang, E., Frolkis, M., Kusler, B., Sage, M., Beeche, M., Bodnar, A.G., Wahl, G.M., Tlsty, T.D., and Chiu, C.P. Telomerase expression in human somatic cells does not induce changes associated with a transformed phenotype. *Nat. Genet.*, *21*: 111–114, 1999.

8. Morales, C.P., Holt, S.E., Ouellette, M., Kaur, K.J., Yan, Y., Wilson, K.S., White, M.A., Wright, W.E., and Shay, J.W. Absence of cancer-associated changes in human fibroblasts immortalized with telomerase. *Nat. Genet.*, *21*: 115–118, 1999.

9. Hahn, W.C., Counter, C.M., Lundberg, A.S., Beijersbergen, R.L., Brooks, M.W., and Weinberg, R.A. Creation of human tumor cells with defined genetic elements. *Nature*, *400*: 464–468, 1999.

10. Wang, J., Hannon, G.J., and Beach, D.H. Risky immortalization by telomerase. *Nature*, *405*: 401–402, 2000.

11. Weitzman, J.B., and Yaniv, M. Rebuilding the road to cancer. *Nature*, *400*: 401–402, 1999.

12. Hahn, C.W., Stewart, S.A., Brooks, M.W., York, S.G., Eaton, E., Kurachi, A., Beijersbergen, R.L., Knoll, J.H.M., Meyerson, M., and Weinberg, R.A. Inhibition of telomerase limits the growth of human cancer cells. *Nat. Med.*, *5*: 1164–1170, 1999.

13. Zhang, X., Mar, V., Zhou, W., Harrington, L., and Robinson, M.O. Telomere shortening and apoptosis in telomerase-inhibited human tumor cells. *Genes Dev.*, *13*: 2388–2399, 1999.

14. Artandi, S.E., Chang, S., Lee, S.L., Alson, S., Golllieb, G.D., Chin, L., and DePinho, R. Telomere dysfunction promotes non-reciprocal translocations and epithelial cancers in mice. *Nature*, *406*: 641–645, 2000.

15. Wright, W.E., and Shay, J.W. Telomere dynamics in cancer progression and prevention: fundamental differences in human and mouse telomere biology. *Nat. Med.*, *6*: 849–851, 2000.

16. Maida, Y., Kyo, S., Kanaya, T., Wang, Z., Yatabe, N., Tanaka, M., Nakamura, M., Ohmichi, M., Gotoh, N., Murakami, S., and Inoue, M. Direct activation of telomerase by EGF through Ets-mediated transactivation of TERT via MAP kinase signaling pathway. *Oncogene*, *21*: 4071–4079, 2002.

17. Sawyers, C.L., McLauglin, J., Goga, A., Havlik, M., and Witte, O. The nuclear tyrosine kinase c-Abl negatively regulates cell growth. *Cell*, *77*: 121–131, 1994.

18. Kharbana, S., Kumar, V., Dhar, S., Pandey, P., Chen, C., Majumder, P., Yuan, Z.-M., Whang, Y., Stauss, W., Pandia, T.K., Weaver, P.D., and Kufe, D. Regulation of the hTERT telomerase catalytic subunit by the c-Abl tyrosine kinase. *Curr. Biol.*, *10*: 568–575, 2000.

19. Vigneri, P., and Wang, J.Y.J. Induction of apoptosis in chronic myelogenous leukemia cells through nuclear entrapment of BCR-ABL tyrosine kinase. *Nat. Med.*, *7*: 228–234, 2001.

20. Baur, J.A., Zou, Y., Shay, J.W., and Wright, W.E. Telomere position effect in human cells. *Science*, *292*: 2075–2077, 2001.

21. Shin-ya, K., Wierzba, K., Matsuo, K., Yamada, Y., Furihata, K., Hayakawa, Y., and Seto, H. Telomestatin, a novel telomerase inhibitor from Streptomyces anulatus. *J. Am. Chem. Soc.*, *123*: 1262–1263, 2001.

Index

Printed in the United States of America